Samuel Leopold Schenk

Elements of Bacteriology for Practitioners and Students,

With Especial Reference to Practical Methods

Samuel Leopold Schenk

Elements of Bacteriology for Practitioners and Students,
With Especial Reference to Practical Methods

ISBN/EAN: 9783744719018

Printed in Europe, USA, Canada, Australia, Japan

Cover: Foto ©berggeist007 / pixelio.de

More available books at **www.hansebooks.com**

ELEMENTS

OF

BACTERIOLOGY

For Practitioners and Students

WITH ESPECIAL REFERENCE TO PRACTICAL METHODS

By Dr S. L. SCHENK

PROFESSOR EXTRAORDINARY IN THE UNIVERSITY OF VIENNA

TRANSLATED FROM THE GERMAN
(BY THE AUTHOR'S PERMISSION)
WITH AN APPENDIX
By W. R. DAWSON, B.A., M.D. Univ. Dubl.

LATE UNIVERSITY TRAVELLING PRIZEMAN IN MEDICINE

WITH 100 ILLUSTRATIONS, PARTLY COLOURED

LONDON
LONGMANS, GREEN, AND CO.
AND NEW YORK: 15 EAST 16th STREET
1893

All rights reserved

TRANSLATOR'S PREFACE

THE present is practically a new edition in English of Professor Schenk's *Grundriss der Bakteriologie*, published in Germany some months ago, the scope and intention of which the Author has explained so fully in his preface that it is unnecessary to touch upon them here. Numerous additions, many of which have been furnished by Professor Schenk, have been made both in the body of the text and in notes. Those by the Translator are distinguished, where of any importance, by square brackets when in the text, as on p. 92, and in the case of foot-notes by square brackets with the letters '—TR.' following, as on p. 21.

The Author originally intended to write an appendix, in order to bring the English edition up to date, but this was found unnecessary, owing to the short time since the appearance of the work in Germany, and the few additions required for this purpose have been incorporated in the text. It was, however, decided at the last moment, in view of the attention lately attracted to those subjects, to furnish a brief account of M. Haffkine's anti-cholera vaccination, together with the results of recent research on the parasitism of protozoa, and on the action of light on bacteria, and this has been done in an appendix, for which the Translator alone is responsible. He has endeavoured to comply with the Author's intention by describing practical details with considerable fulness, but has thought it best to depart from the rule of not giving references, the observations on these subjects being so recent that certainty has not yet been attained. These remarks apply also to the note at the end of Chapter X. on Sabouraud's recent researches. A few additional methods and formulæ have also been given, the body of the work has been divided into chapters, and the index greatly extended.

The Translator has to express his thanks for assistance on individual points from various sources, but especially to Professor Schenk for his unvarying courtesy and readiness in affording additional information; and to acknowledge occasional obligations to many of the leading English and German text-books.

W. R. DAWSON.

Dublin: *June* 1893.

AUTHOR'S PREFACE

THE present work is intended to furnish the student and the practitioner with a guide to the science of Bacteriology ; consequently the methods of investigation have been dealt with as thoroughly as possible, special attention being paid to the elementary technique. The Author has thought it best to describe the micro-organisms according to the localities in which they are met with, a plan which rendered it possible to go thoroughly into the respective methods of research in their proper places. Particular attention has been given to the pathogenic micro-organisms, and so much the more regard had to be paid to the chemical relations of the life of bacteria, and to their biology generally, that recent events give us reason to hope for an extension of our therapeutic powers from this direction.

Engravings of the most important bacteria have been provided, showing their form and the appearances presented by their growth. These are intended to serve the reader as models of the typical forms, to which he may be able to adhere in his own investigations.

The Author has endeavoured in the text to consider the views of all the different schools, and to this end he has consulted other manuals and bacteriological publications. Conformably to the scope of a handbook like the present, however, all references to the literature have been omitted.[1]

It is his hope that this manual may contribute to preserve and promote the interest felt by practitioners and students in the science of Bacteriology.

S. L. SCHENK.

[1] The words omitted have reference to the German edition.

CONTENTS

CHAPTER I
GENERAL MORPHOLOGY AND BIOLOGY OF MICRO-ORGANISMS

PAGE

INTRODUCTORY—Varieties of Micro-organisms—Bacteria—Motility of Bacteria—Capsules of Bacteria—Multiplication of Bacteria—Products of Metabolism in Bacteria—Influence of Bacteria on the Tissues—Toxines, Toxalbumins, and Ptomaines—Moulds—Yeasts —Algæ—Protozoa—Examination of Micro-organisms .   1

CHAPTER II
PRELIMINARY PROCESSES—APPARATUS AND REAGENTS

STERILISATION—Sterilisation by Heat—By Steam—Fractional Sterilisation—Sterilisation by Steam under Pressure—Chemical Disinfectants. APPARATUS AND REAGENTS—Microscope—Steam Steriliser —Incubator—Thermo-regulator— Schenk's Thermo-regulator— Meyer's Thermo-regulator—Gärtner's—Altmann's—Petroleum Incubator—Miscellaneous Apparatus—Hot-water Filter—Apparatus for Plate Cultivations—Moist Chambers—Plates—Petri's Capsules —Soyka's Plates—Wire Crates—Reagents—Stains—Other Utensils —Centrifugal Machines   15

CHAPTER III
NUTRIENT MATERIALS AND METHODS OF CULTIVATION

NUTRIENT MEDIA—Liquid Nutrient Media—Preparation of Meat Bouillon—Preparation of Meat-extract Bouillon—Solutions of White of Egg—Solid Nutrient Media—Preparation of Peptone Bouillon Gelatine—Of Meat-extract Peptone Gelatine—Additions to Nutrient Gelatine—Preparation of Urine Gelatine—Preparation of Nutrient Agar—Of Peptone Bouillon Agar—Modifications of

Gelatine and Agar, &c.—Blood Serum—Modifications of Serum—
Eggs of Birds—Plovers' Egg Albumen—Hens' Eggs—Potatoes—
Rice, Bread, and Wafers. MODES OF CULTIVATION—Slide Cultures
—Koch's Plate Process—Roll Cultures—Modifications of the Plate
Process—Plate Cultures on Serum and Plovers' Egg Albumen—
Cultivation of Anaerobic Micro-organisms—Permanent Cultures   35

CHAPTER IV

EXAMINATION OF MICRO-ORGANISMS UNDER THE MICROSCOPE AND BY EXPERIMENTS ON LIVING ANIMALS

EXAMINATION in the Fresh State—In the Hanging Drop—Staining of Micro-organisms—Simple Staining of Cover-glass Preparations—Preparation of Stain-Solutions—Staining of Flagella—Of Spores—DECOLORISING AGENTS—Koch and Ehrlich Method of Staining—Ziehl and Neelsen's Method—Ehrlich's—Günther's—Weichselbaum's—Fraenkel's Gabbet's—Method of Pfuhl and Petri—Method of Pittion Arens's Chloroform Method Gram's Decolorising Method Günther's Modification of Gram's Process—Weigert's Modification of Gram's Process—Impression Preparations—Examination of Micro-organisms in Sections of Tissue—Examination by the Freezing Method—Hardening Imbedding—Imbedding in Gum Arabic—In Glycerine Jelly—In Celloidine—In Paraffine. ON STAINING OF SECTIONS—Unna's Drying-on Process—Combination of Staining Methods Kühne's Methyl Blue Method—Koch's Method—Löffler's—Chenzynsky's—Gram's—Kühne's Modification of Gram's—Kühne's Dry Method—Weigert's Iodine Method—Unna's Borax Methyl Blue Method—Unna's Methods of Demonstrating the Organisms of the Skin—Noniewicz's Method. EXPERIMENTS ON LIVING ANIMALS—Transmission of Micro-organisms to Animals—Infection by the Air-passages—Infection by the Digestive Canal—Subcutaneous Infection—Intravenous Infection—Infection into the Anterior Chamber of the Eye 65

CHAPTER V

THE BACTERIOLOGICAL ANALYSIS OF AIR

MICRO-ORGANISMS IN THE AIR—Simple Methods of Examining Air—Pouchet's Method—Miquel's—Emmerich's—Welz's—Hesse's—Method of Strauss and Wurz—Petri's Method—Tyndall's Method—Penicillium glaucum—Brown Mould—Yeast—Micrococcus radiatus—Micrococcus versicolor—Micrococcus cinabareus—Micrococcus flavus tardigradus—Micrococcus candicans—Micrococcus viticulosus—Micrococcus Ureæ—Micrococcus roseus—Diplococcus citreus conglomeratus—Micrococcus flavus liquefaciens and

CONTENTS

Micrococcus desidens—Sarcina alba—Sarcina candida—Sarcina aurantiaca — Sarcina rosea — Sarcina lutea — Staphylococci — Staphylococcus pyogenes aureus; albus; citreus—Streptococci—Bacillus subtilis—Bacillus prodigiosus—Potato Bacillus—Bacillus mesentericus fuscus; ruber; vulgatus — Bacillus liodermos — Bacillus melochloros—Bacillus multipediculosus—Bacillus neapolitanus—Atmospheric Spirilla 96

CHAPTER VI

THE BACTERIOLOGICAL ANALYSIS OF WATER

MICRO-ORGANISMS OF WATER—Filtration and Filters—Variations in Water Depending on Source—Examination of Water—Pfuhl's Method Kirchner's Method—Other Methods—Micrococcus aquatilis—Micrococcus agilis—Micrococcus fuscus—Micrococcus luteus —Micrococcus aurantiacus—Micrococcus fervidosus—Micrococcus carneus—Micrococcus concentricus—Diplococcus luteus—Bacillus fluorescens liquefaciens and Bacillus nivalis (Glacier Bacillus) —Bacillus fluorescens non-liquefaciens—Bacillus Erythrosporus—Bacillus arborescens—Bacillus violaceus—Bacillus gasoformans—Bacillus phosphorescens (indigenus; indicus)—Bacillus ramosus —Bacillus aurantiacus—Bacillus aureus—Bacillus bruneus—Bacillus aquatilis—Bacillus aquatilis sulcatus—Bacillus aquatilis radiatus—Bacterium Zürnianum—Bacillus membranaceus amethystinus — Bacillus indigoferus — Bacillus ianthinus — Bacillus ochraceus—Bacillus gracilis—Bacillus sulphydrogenus—Bacillus of Asiatic Cholera and Allied Micro-organisms—Vibrio proteus—Vibrio Metschnikoffi—Bacillus of Typhoid Fever—Bacterium Coli commune—Spirilla in Water—Other Micro-organisms of Water . 122

CHAPTER VII

BACTERIOLOGICAL ANALYSIS OF EARTH AND OF PUTREFYING SUBSTANCES

MICRO-ORGANISMS IN THE SOIL—Method of Examination—Bacillus mycoides (Earth Bacillus) Bacterium mycoides roseum—Bacillus radiatus — Bacillus spinosus — Bacillus liquefaciens magnus—Bacillus scissus—Clostridium foetidum—Bacillus œdematis maligni—Bacillus of Tetanus—Streptococcus septicus—Bacillus Anthracis—Plasmodium Malariæ—Other Micro-organisms of the Soil. ANALYSIS OF PUTREFYING SUBSTANCES—Differences in Putrefactive Processes—Bacillus fuscus limbatus—Proteus—Proteus vulgaris — Zenkeri — Mirabilis — Hominis—Capsulatus — Bacillus saprogenes—Spirillum concentricum—Spirillum rubrum . . 155

CHAPTER VIII

MICRO-ORGANISMS IN ARTICLES OF DIET

Methods of Examining Different Foods. EXAMINATION OF MILK— Methods —Bacillus lacticus (Bacillus acidi lactici)—Micrococcus acidi lactici — Clostridium butyricum (Bacillus amylobacter) — Micrococcus acidi lactici liquefaciens—Oidium Lactis—Bacillus butyricus—Bacillus Butyri viscosus; fluorescens—Spirillum tyrogenum—Bacillus Lactis viscosus—Bacillus Lactis pituitosi—Bacillus actinobacter—Bacillus fœtidus Lactis—Bacillus cyanogenus—Bacterium Lactis erythrogenes—Sarcina rosea—Micrococcus of Bovine Mastitis —Other Pathogenic Bacteria in Milk—Saccharomyces ruber—Bacillus caucasicus (Dispora caucasica, or Kephir Bacillus). EXAMINATION OF OTHER ARTICLES OF DIET—Bacillus megaterium— Bacillus Aceti—Bacillus indigogenus—Pediococcus Cerevisiæ—Sarcina Cerevisiæ—Micrococcus viscosus—Bacillus viscosus Cerevisiæ—Bacillus viscosus Sacchari—Moulds on Articles of Food—Aspergillus niger; albus; glaucus; flavescens; fumigatus—Mucor Mucedo; rhizopodiformis; corymbifer; ramosus 178

CHAPTER IX

BACTERIOLOGIAL EXAMINATION OF PUS

Properties and Composition of Pus—Actinomyces—Bacillus pyocyaneus α and β—Staphylococcus cereus albus; flavus; aureus—Streptococcus pyogenes—Micrococcus of Gonorrhœa - Bacillus of Syphilis — Bacillus Tuberculosis — Bacillus of Glanders — Other Microbes of Pus 196

CHAPTER X

BACTERIOLOGICAL EXAMINATION OF THE ORGANS AND CAVITIES OF THE BODY AND THEIR CONTENTS

MICRO-ORGANISMS OF THE LIVING BODY - I. THE SKIN—Micro-organisms of the Skin -Methods of Examination—Diplococcus subflavus—Micrococcus lacteus faviformis—Diplococcus albicans amplus—Diplococcus albicans tardus—Diplococcus citreus liquefaciens—Diplococcus flavus liquefaciens tardus—Micrococcus hæmatodes —Micrococcus of Trachoma —Diplococcus of Acute Pemphigus—Vaginal Bacillus —Bacillus of Symptomatic Anthrax— Lepra Bacillus Bacillus sycosiferus fœtidus—Ascobacillus citreus—Bacillus Xerosis—Trichophyton tonsurans—Fungus of Favus (Achorion Schœnleinii)—Microsporon furfur. NOTE: Trichophyton microsporon - Macrosporon 222

CONTENTS xiii

CHAPTER XI

THE ORGANS AND CAVITIES OF THE BODY AND THEIR CONTENTS—(*cont.*) II. THE DIGESTIVE TRACT

PAGE

THE CAVITY OF THE MOUTH—Micro-organisms of the Mouth and their Examination—Leptothrix—Bacillus buccalis maximus—Iodococcus —Micrococcus salivarius septicus and Bacillus salivarius septicus— Bacillus ulna—Bacillus Gingivæ—Bacillus Diphtheriæ—Spirillum Miller—Spirochæte Dentium (Denticola)—Vibrio rugula—Fungus of Thrush—Other Bacteria of the Mouth. THE TYMPANUM. THE STOMACH—Micro-organisms of the Stomach—Sarcina Ventriculi— Micrococcus tetragenus mobilis Ventriculi—Bacterium Lactis aerogenes—Bacillus indicus—THE INTESTINE—Intestinal Micro- organisms—Micrococcus aerogenes—Bacillus putrificus Coli— Bacillus coprogenes fœtidus—Bacterium Zopfi—Bacterium aero- genes, Helicobacterium aerogenes, and Bacillus aerogenes—Bacillus Dysenteriæ—Bacillus of Fowl Cholera—Other Intestinal Bacteria . 237

CHAPTER XII

THE ORGANS AND CAVITIES OF THE BODY AND THEIR CONTENTS—(*cont.*) III. THE FÆCES AND URINE

THE FÆCES—Composition and Modes of Examining—Bacillus subti- liformis—Bacillus albuminis—Bacillus cavicida—Micrococcus tetragenus concentricus. THE URINE—Micro-organisms of the Urine—Yeasts and Moulds in Urine—Urobacteria - Staphylococcus Ureæ candidus Liquefaciens—Micrococcus Ureæ liquefaciens— Bacilius Ureæ Urobacillus Freudenreichii—Madoxii Micrococ- cus ochroleucus—Streptococcus giganteus Urethræ—Bacterium sulphureum—Bacillus septicus Vesicæ—Urobacillus liquefaciens . 254

CHAPTER XIII

THE ORGANS AND CAVITIES OF THE BODY AND THEIR CONTENTS—(*cont*). IV. BACTERIOLOGICAL EXAMI- NATION OF THE RESPIRATORY TRACT, AND (V.) OF THE BLOOD

THE NASAL SECRETION—Micro-organisms in the Nasal Secretion— Micrococcus cumulatus tenuis—Micrococcus tetragenus subflavus— Micrococcus Nasalis—Diplococcus Coryzæ—Staphylococcus cereus aureus—Bacillus fœtidus ozænæ—Bacillus striatus albus et flavus —Bacillus of Rhinoscleroma—Bacillus capsulatus mucosus—Vibrio Nasalis—Other Nasal Bacteria. THE RESPIRATORY PASSAGES— Micro-organisms of the Respiratory Passages- Sarcina Pulmonum

BACTERIOLOGY

— Sarcina aurea — Diplococcus Pneumoniæ — Pneumobacillus Friedlænderi—Micrococcus tetragenus—Bacillus aureus—Tubercle Bacillus and Actinomyces—Bacillus Tussis convulsivæ—Bacillus pneumosepticus. V. BACTERIOLOGICAL EXAMINATION OF THE BLOOD— Micro-organisms in the Blood—Methods of Examination—Influenza Bacillus — Bacillus Endocarditis capsulatus — Bacillus of Swine Erysipelas—Bacillus murisepticus—Spirochæte Obermeieri —Protozoa in the Blood 264

APPENDIX (by the Translator)

A. VACCINATION AGAINST ASIATIC CHOLERA—Principle of Anti-cholera Vaccination—Preparation of Vaccine—Results of Vaccination. B. PARASITIC PROTOZOA — Pathogenesis in Protozoa—Coccidium oviforme—Amœba Dysenteriæ—Protozoa in Carcinoma—Protozoa in other New Growths, &c. C. THE ACTION OF LIGHT ON MICRO-ORGANISMS—Action of White Light—Action of Coloured Light— Mode of Action of Light—Applications in Nature. D. ADDITIONAL METHODS AND FORMULÆ—Fixing Methods—The Gum Freezing Method—Staining Formulæ 283

INDEX 303

BACTERIOLOGY

Errata

Page 11, line 5 from bottom, *for* CRENOTHLE *read* CRENOTHRIX

,, 236, line 4, the words NOTE BY TRANSLATOR refer to what follows, not to what precedes.

,, 283 line 4, from bottom (note 1), *for* Klemperei *read* Klemperer.

,, 298, note 1, *for* quoted without reference, &c., *read* *Pflüger's Archiv*, xxx, p. 95.

,, 302, lines 5-3 from bottom should read as follows: minutes in Löffler's or Kühne's methyl blue, washed, dipped for an instant in 10 per cent. tannin solution, slightly decolorised in feebly acid water, washed in water,

and *facultative* parasites—the obligate or strict parasites being those which grow exclusively in the living body and perish apart from it, whereas the facultative parasites have the power of adapting themselves to altered conditions of life, and of flourishing externally as well as internally. With further reference to their relation to the human or animal frame we speak also of *ectogenous* micro-organisms, which occur outside the body, of *endogenous*, which exist in its interior, and of *ambigenous*, which are capable of life either within or without. Again, the majority are unable to live without oxygen, and these are termed *aerobes*; but a large number

B

— Sarcina aurea — Diplococcus Pneumoniæ — Pneumobacillus Friedlænderi—Micrococcus tetragenus—Bacillus aureus—Tubercle Bacillus and Actinomyces—Bacillus Tussis convulsivæ—Bacillus pneumosepticus. V. BACTERIOLOGICAL EXAMINATION OF THE BLOOD— Micro-organisms in the Blood—Methods of Examination—Influenza Bacillus — Bacillus Endocarditis capsulatus — Bacillus of Swine Erysipelas—Bacillus murisepticus—Spirochæte Obermeieri —Protozoa in the Blood 264

APPENDIX (*by the Translator*)

A. VACCINATION AGAINST ASIATIC CHOLERA—Principle of Anti-cholera Vaccination—Preparation of Vaccine—Results of Vaccination.

BACTERIOLOGY

CHAPTER I

GENERAL MORPHOLOGY AND BIOLOGY OF MICRO-ORGANISMS

Introductory.—Varieties of micro-organisms.—Within and without the human body there exist countless organisms of microscopic minuteness, *micro-organisms*, which, alighting upon the surface, may effect an entrance in various ways into the interior; they belong partly to the vegetable, partly to the animal kingdom, and have the property of developing on animal and vegetable bodies. According as they are capable of growth upon dead substances or living matter, we distinguish respectively *saprophytes* and *parasites*; the latter of which are subdivided into *obligate* and *facultative* parasites—the obligate or strict parasites being those which grow exclusively in the living body and perish apart from it, whereas the facultative parasites have the power of adapting themselves to altered conditions of life, and of flourishing externally as well as internally. With further reference to their relation to the human or animal frame we speak also of *ectogenous* micro-organisms, which occur outside the body, of *endogenous*, which exist in its interior, and of *ambigenous*, which are capable of life either within or without. Again, the majority are unable to live without oxygen, and these are termed *aerobes*; but a large number can thrive whether it is absent or not and are called *facultative anaerobes*; while those micro-organisms whose growth can make no progress in the presence of the gas are denominated *obligate anaerobes*.

Micro-organisms belong to the following different classes—viz. *bacteria, moulds, yeasts, algæ, and protozoa*.

Bacteria, schizomycetes, or fission-fungi, are colourless cells of a glassy transparency, possessing an enveloping membrane with protoplasmic contents but no nucleus, and having a length which amounts in general to a few thousandths, and a breadth of some ten-thousandths, of a millimetre. The interior of the bacterial cell has usually a homogeneous appearance, but sometimes shows oil-like granules. Bacteria are distinguished according to their form as *cocci, bacilli*, and *spirilla*.

Cocci are globular in shape, and are found either singly or united in groups. If they lie singly they are called *monococci*; if grouped in masses, *staphylococci*; if the elements are joined in pairs and fours we distinguish respectively, according to the number, *diplococci* and *tetracocci*; if each eight is so united as to resemble a bale of goods, they are named *sarcinæ*; and if they are strung together in chains, *streptococci*.

Bacilli are minute straight rods, the smallest discovered up to the present time being the influenza bacillus. Their ends are sometimes sharply cut across, sometimes rounded off, and the rodlets themselves are in some cases thin, in others stout and thick, in others again swollen in the centre, and so forth.

Spirilla are spirally curved rods, and are subdivided into comma-bacilli or *vibrios, spirilla* in the more restricted sense, and *spirochætæ*. The vibrios usually form strings of cells which strongly resemble spirilla; the spirochætæ are distinguished by their flexibility.

Motility of bacteria.—A large number of bacteria possess the power of movement, accomplished by means of *cilia* or *flagella*, forming spiral processes, of which sometimes one is present, sometimes several, and which keep up an incessant vibration.

This *motility* must not be confounded with oscillatory movements (also, however, to be understood as originated by

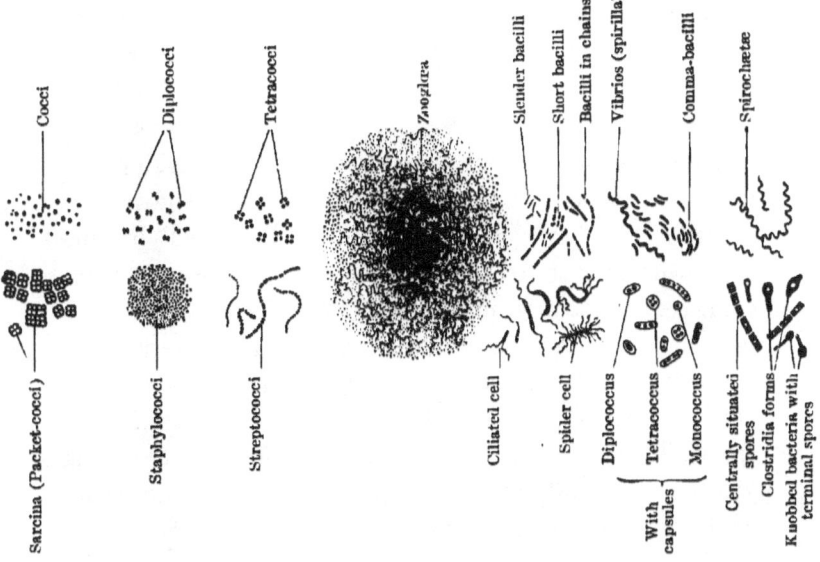

FIG. 1.—FORMS OF BACTERIA. Magnified about 700 times. (After Baumgarten.)

the bacteria themselves), nor with such motion as occurs in inert particles suspended in a fluid medium, and which is known as the 'molecular movement of Brown' (*Brownian movement*).

Capsules of bacteria.—Capsules are formed by swelling of the membrane, and are best seen in those microorganisms which are found in groups—viz. the diplococci, streptococci, tetracocci, and sarcinæ; but various bacilli also, such as Pfeiffer's *capsule-bacillus* and the *Bacillus* *capsulatus mucosus* of Fasching, are furnished with capsules. By coalescence of the cell-capsules conglomerations of cells are formed which are called *zoogloea*, and may spread over the surface of fluids in the form of a *pellicle*. This pellicle sometimes serves to distinguish between micro-organisms which strongly resemble each other—for instance, cholera bacilli form a pellicle, whereas Finkler-Prior bacilli do not.

Multiplication of bacteria takes place by *fission*, hence their name of *Schizomycetes*, or ' fission-fungi.' As soon as the individual organism has attained its normal size, there appears in the centre a clear line which forms the sign of division, the two individuals so produced then breaking free from one another and forming again independent organisms. If, however, the daughter-cells do not become disjoined, groups are formed in which the cells remain connected in strings, in clusters (staphylococci), in chains (streptococci), and so on, the spiral strings formed by the vibriones being often wrongly described as spirilla. If the division of cocci takes place in one direction of space, diplococci are formed; division in two directions yields as a result tablet-cocci (merismopedia, tetracocci); while division in three directions gives packet-cocci or sarcinæ.

A second mode in which bacteria propagate is multiplication by the development of *spores*. These are distinguished by their very remarkable power of resisting the influences of temperature and the action of chemicals, and are therefore called also *permanent* forms. The spores in most cases occupy a position in the centre of the bacterial cell, but in a few varieties they are at the end. Sometimes they cause a bulging of the centre of the cell, so that the latter becomes spindle-shaped, a form which is known as *clostridium* (see fig. 1). In the cases in which they occupy one end, as in the *Tetanus bacillus*, the cells show some resemblance to a drum-stick.

Formation of spores in the interior of the mother-cell is described as *proper* or *endogenous*, while *arthrogenous* is the term applied to that which takes place when single portions separate from the cell and develop gradually into independent individuals (*arthrospores*).

Bacteria require for their growth a certain amount of moisture, many of them speedily perishing if dried.

Products of metabolism in bacteria.—The biological properties of bacteria are, next to their morphological peculiarities, of special importance.

A large number of micro-organisms have the property of generating colouring matter, though not chlorophyll. The bacteria are themselves colourless and transparent, and the pigment is merely formed as a product of their metabolism, especially under the influence of light.

Many bacteria throw off odorous products, and some anaerobic micro-organisms generate very foul putrefactive gases (ammonia, sulphuretted hydrogen, scatol, &c.) The bacillus of Asiatic cholera exhales a pleasantly aromatic odour, and the *Bacillus prodigiosus* a smell resembling that of trimethylamine.

Micro-organisms have also the property of producing changes in the medium on which they are cultivated by the products of their metabolism, so that albuminoid substances are peptonised by some of them and gelatine liquefied. Many have the faculty of resolving organic bodies into their simplest elements; but some possess the power of forming nitrates by the conversion of ammonia into nitrous and nitric acids. Certain microbes break up the chemical combination of albuminoid bodies, causing putrefaction, while others induce fermentation; and a small number, again, have the property of becoming luminous in the dark (phosphorescence), in consequence of the molecular activity of their protoplasm.

Influence of bacteria on the tissues.—The influence which bacteria exert on the tissue of the bodies of men and animals depends both on the qualities of the bacteria and on the nature of the tissue. The action of the pathogenic bacteria is not alike in all animals, and those which are insusceptible to certain bacteria are said to be *immune* in their relation to those particular organisms. Animals, however, which are commonly immune towards a pathogenic micro-organism may, under altered conditions, lose their immunity; for example, the frog, although usually immune to anthrax, becomes susceptible to it at a higher temperature.

The virulence of pathogenic bacteria may be diminished by various factors, amongst which are included sojourn in the body of immune animals, increased atmospheric pressure, the action of higher degrees of temperature, the influence of sunlight, &c. This weakening is dependent on an alteration in the products of metabolism.

Toxins, toxalbumins, and ptomains.—As products of the metabolism of those bacteria which effect an entrance into the body substances are formed, some of which exert violent toxic action, and which are divided into *toxins* and *ptomains*, or cadaveric alkaloids, and to these is ascribed the action of the pathogenic bacteria in originating morbid processes. The toxins are further subdivided into *toxalbumins and proteins*.

Toxalbumins are albuminoid bodies formed during the growth of the bacteria upon culture-media, especially in bouillon. Their activity depends upon certain definite degrees of temperature, and they decompose at the boiling-point. Our knowledge of them is due to the researches of Roux, Yersin, Brieger, and C. Fraenkel.

Proteins are albuminoid bodies which are contained in the actual substance of the bacteria, and whose chief point of difference from toxalbumins consists in their not being decomposed by boiling, even if kept up for hours. They were discovered by Nencki, and further studied by Buchner. Proteins can be abundantly obtained by boiling pure cultures of bacteria in their bouillon or mixed with water. Koch's *tuberculin* belongs to this group of substances.

Toxins, when injected in small quantity into animals, have the property of affording immunity from infection with the corresponding bacterium.

According to Brieger, the cultures or juice from the tissue should be filtered by means of earthenware cells, so as to free the liquid from living germs. The albuminoid substances are then precipitated with ten times the quantity of absolute alcohol, redissolved in dilute alcohol, and precipitated a second time with an alcoholic solution of corrosive sublimate. The mercury is next removed by means of sulphuretted hydrogen, the residue dissolved in water, and again treated with sulphuretted hydrogen; this process having been several times repeated, the toxins are finally precipitated from the aqueous solution by absolute alcohol.

Scholl produced a *toxopeptone* from cultures of cholera-bacilli in hens' eggs, without filtration, by the following process:—The albumen liquefied by the bacteria was poured into ten times its quantity of absolute alcohol, and the precipitate washed with alcohol, digested with water, and filtered. The aqueous solution was then repeatedly added to ether and alcohol slightly acidulated with acetic acid, decanted each time from the residue, and the latter redissolved in water rendered alkaline, after which a final addition to pure ether, which was then evaporated off, resulted in the obtaining of the poisonous substance.

To procure ptomains from the actual bacterial cells

after the method of Buchner potato cultures are prepared, and the mass of bacteria is then scraped from them, and rubbed up in a mortar with a little water, mixed with fifty times its volume of a half per cent. solution of caustic potash, and then digested in a water-bath until the greatest possible degree of fluidity is reached, after which it is filtered through several small filters. Dilute acetic or hydrochloric acid is next added to the filtrate until, while avoiding any excess of acidity, the reaction becomes distinctly acid. The protein precipitated in this way is collected on a filter, washed, and dissolved in water which is feebly alkaline.

Roemer procured his extracts of the *Bacillus pyocyaneus* and *pneumobacillus* in the following manner :—The masses of bacteria are carefully scraped from well-developed cultures on potato and rubbed to a fine emulsion with ten times their bulk of distilled water. The emulsion, having been sterilised by boiling for several hours, is left for about four weeks in the incubator, during which time it must frequently be boiled for an hour or two, so that in the course of the process it is boiled for from thirty to forty hours in all. When the four weeks have expired the emulsion is filtered through a tubular filter made of kaolin (*Chamberland's candle*), or through one of infusorial earth, and the resulting filtrate is a clear brownish or yellowish fluid containing albuminoid substances.

Koch obtained his *tuberculin* by extraction from pure cultures of tubercle bacilli, which he grew upon a feebly alkaline infusion of veal containing an addition of one per cent. peptone and four to five per cent. glycerine. The culture-vessels are inoculated by floating a fairly large piece of the seed-culture on the surface of the fluid, and are then kept at a temperature of 38° C. In from three to four weeks the surface is covered with a tolerably thick membrane, dry above, and often thrown into folds, which in two or three weeks more becomes moistened by the fluid, and finally breaks up into ragged pieces and sinks to the

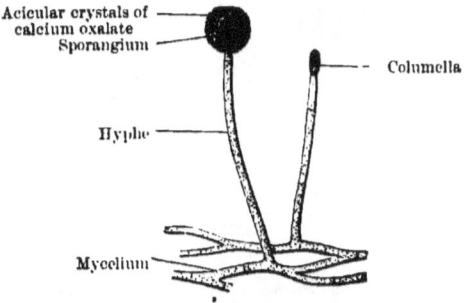

FIG. 2.—MUCOR MUCEDO. (After Baumgarten.)

bottom. The cultures (which thus require from six to eight weeks for their growth), when fully ripe, are evaporated to a tenth of their bulk in a water-bath, and filtered through earthenware or infusorial earth.

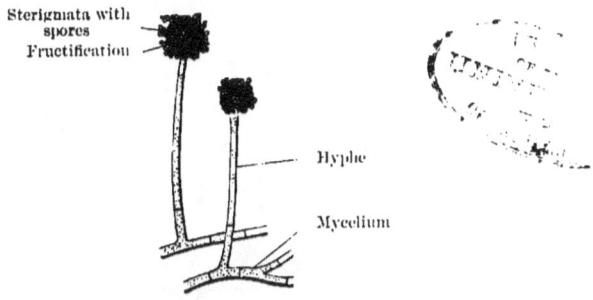

FIG. 3.—ASPERGILLUS GLAUCUS.

Moulds.—These are for the most part saprophytes, though pathogenic varieties are also to be found among them. They form spores which, like those of bacteria, are marked by the strong resistance they offer to external influences, and which develop under favourable circumstances into complete individuals. They sometimes contain shining drops like fat-globules.

A tubular bud pushes out from the enveloping membrane of the spore, lengthens by growing at the end, and quickly forms a very freely-branching network of fibres,

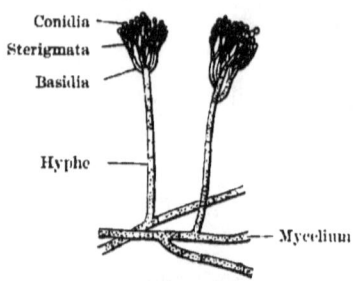

FIG. 4.—PENICILIUM GLAUCUM. Magnified 400 times. (After Baumgarten.)

spoken of as *mycelium*, and possessing special seed-bearing organs called *hyphæ* or *thallus*, from which the moulds derive the name of *Hyphomycetæ*. According to the form of the seed-organ they are divided into *mucorineæ*, *aspergillineæ*, *penicilliaceæ*, and *oidiaceæ*.

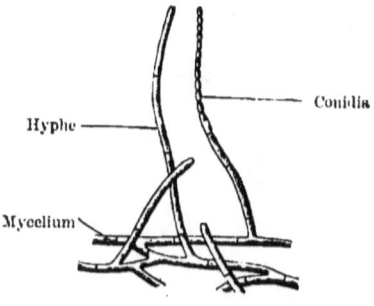

FIG. 5.—OIDIUM LACTIS. (After Baumgarten.)

In the *mucorineæ*, or headed moulds, the ends of the hyphæ swell into knobs (*columella*), around which a seed-capsule, or *sporangium*, forms. In this the spores develop in such a way as to burst the enveloping epicarp membrane when fully ripe (fig. 2).

The *aspergillineæ* (knob-moulds) have the knobbed ends

of the hyphæ covered with a variable number of spore-carriers, or *sterigmata*, from the extremities of which the spores divide off in rows (fig. 3).

The hyphæ of the *penicilliaceæ* (pencil-moulds) are branched, which is not the case with the mucor and asper-

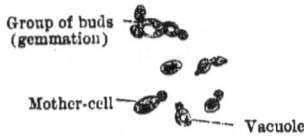

FIG. 6.—YEAST CELLS (*Saccharomyces Cerevisiæ*). Magnified 900 times.

gillus varieties, and on the terminal twigs of the tuft so formed (the *basidia*) are seen the sterigmata, from which the spores, or *conidia*, are separated off in the form of chains (fig. 4).

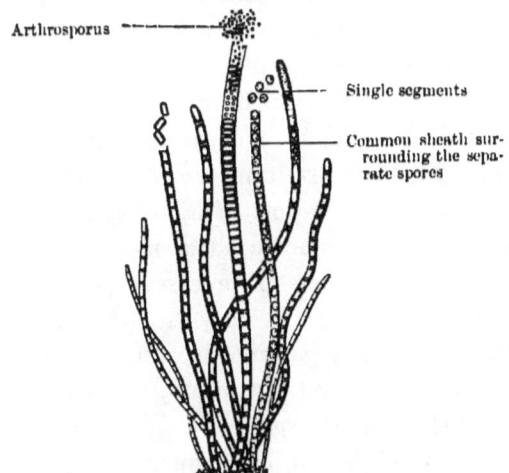

FIG. 7.—CRENOTHLÆ KUHNIANA. Magnified 600 times. (After Zopf.)

The *oidiaceæ* are distinguished by the fact that the hyphæ form no special spore-bearing organs, but become articulated at their extremities, and so divide off the spores in the form of segments.

Yeasts.—These possess neither spore-bearing organs nor spores, but multiply by *gemmation*, which consists in the budding out of daughter-cells in different places from the gradually enlarging mother-cell, these in their turn becoming mother-cells, thus forming groups of buds. The individual yeast-cells are round or elliptical, and often display in their interior colourless lacunæ, which are not spores, but may perhaps consist of minute drops of fat, and are called *vacuoles*. The yeasts play an important part in nature in causing fermentation. Several species of them form pigments.

Algæ.—Of these the *cladothrix*, *crenothrix*, and *beggiatoa* varieties belong to the micro-organisms. They are jointed filaments, which multiply not by fission but by germination at their extremities (fig. 7).

Protozoa.—Of the protozoa those important as regards bacteriological investigation are the *sporozoa*, which include the *gregarinæ*, *psorospermii*, and *coccidia*. They are unicellular organisms which can only live in a moist or liquid medium, and in the absence of water, nutrient material, or oxygen, are transformed into roundish durable *cysts*. They possess a sort of larval condition, consisting of irregular and roundish little masses of protoplasm, which move by means of processes projecting out like limbs (*pseudopodia*), or by flagella, and often, losing their mobility, take up a permanent residence in other cells. The contents of the cyst separate by division or gemmation into particles called *sporocysts* or *pseudonavicella*, the contents of which, again, break up into a number of sickle-shaped germs. Pfeiffer considers the plasmodia of malaria to be also cysts of this nature containing crowds of spores.

Examination of micro-organisms.—Microscopic examination alone is not sufficient to establish fully the properties of micro-organisms in their morphological and biological relations. The attempt must be made to obtain pure cultivations, which are then, on the one hand, to be sub-

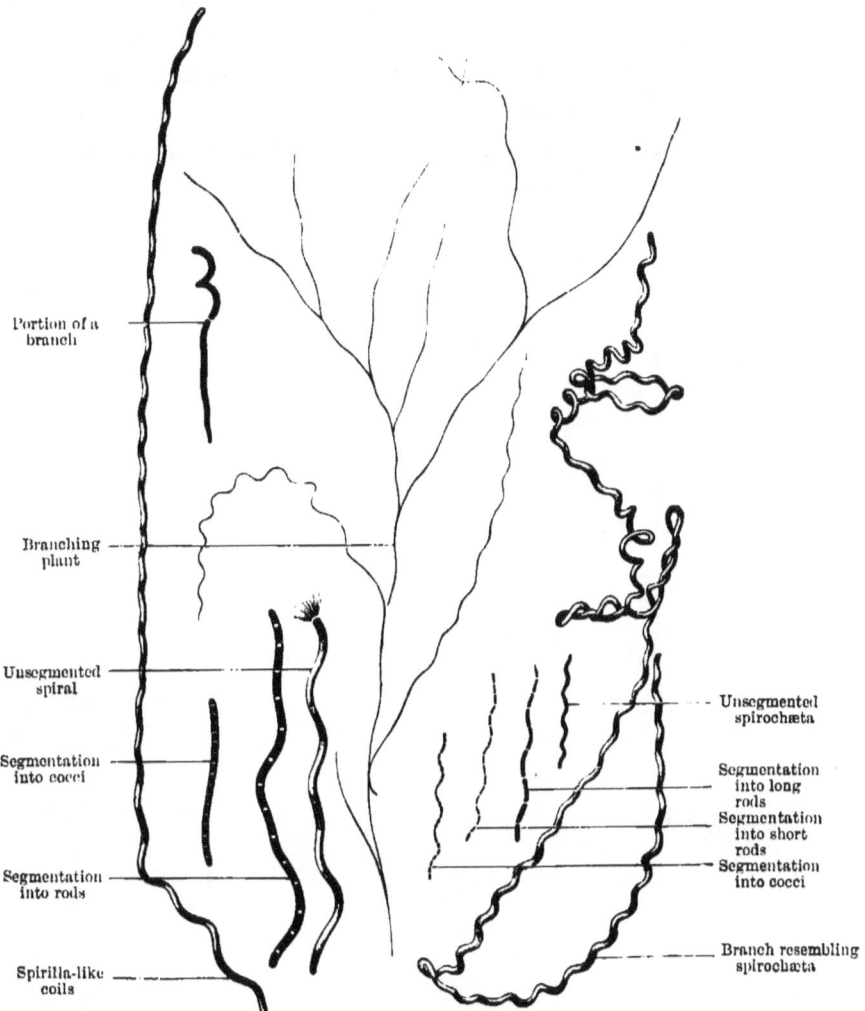

FIG. 8.—FORMS OF VEGETATION OF CLADOTHRIX DICHOTOMA. (After Zopf.)

mitted to microscopic examination, and on the other transferred to substances liable to fermentation and putrefaction, and used for experiments on animals. In order to procure pure cultures, however, the instruments and utensils employed must be freed from the micro-organisms adhering to them, or *sterilised*. It seems, therefore, advisable first to discuss the methods of sterilisation, and then the preparation of culture-media, passing on afterwards to microscopic examination, and, finally, the methods of transmission to living animals.

CHAPTER II

PRELIMINARY PROCESSES—APPARATUS AND REAGENTS

Sterilisation is the process by which both instruments and culture-media are freed from living germs, and is carried out in different ways.

Sterilisation by heat.—Articles capable of withstanding a very high temperature are best sterilised by being held in the flame of a Bunsen burner or spirit-lamp until a red or white heat is reached, a method which is especially applicable to small platinum wires or plates.

Bodies which have no great power of resistance, and instruments which could not be exposed to so intense a heat without impairing their efficiency and sharpness, are subjected for a longer time to a temperature of 150° C., by which both the micro-organisms and also their spores are destroyed. For this purpose a box made of sheet iron, with double walls, is best employed, which has thus a layer of air between the walls (*hot-air steriliser*). It must be put together with rivets, not with solder. By means of a single powerful burner or a number of small ones placed beneath, the temperature of the interior is rapidly brought to 160°-170°, after which half an hour suffices for sterilisation. In the top of the box are openings for two thermometers, one of which extends into the interior, while the other registers the temperature of the space between the inner and outer walls; and there is also a valve for the purpose of regulating the temperature. This arrangement is suited for the sterilisation of instruments, apparatus, glassware, &c. (fig. 9).

To prevent flasks and test-glasses from becoming re-infected after sterilisation they must be closed with a plug of cotton-wool before being placed in the steriliser, as such a plug, while allowing the air to enter the vessels, keeps back the organisms floating in it. The plug can be further covered with a cap of indiarubber.

Instead of plugging with cotton, Staff-surgeon Schill recommends the use of *double test-glasses*, consisting of two test-tubes made of stout glass and with smooth even edges, one of which is pushed over the other as a cover. The former should be only so much wider as to leave a space the thickness of a sheet of paper between the two, and should be but half as long as the lower.

FIG. 9.—HOT-AIR STERILISER.

Sterilisation by steam.—Articles which cannot be exposed to so high a temperature are sterilised by the vapour of boiling water at 100° C. For this purpose a cylindrical vessel of sheet copper is used, measuring nearly a metre in height and about 20 cm. in diameter, covered with felt or asbestos to prevent loss of heat, and capable of being closed with a lid similarly protected. The latter is provided with an opening for the introduction of a thermometer, and does not fit quite air-tight (fig. 10). The bottom of the vessel is double, the inner bottom consisting of a grating fixed about 30 cm. above the outer, and the space between the two is about half filled with water, the height of which is observed by a gauge-tube at the side (*Koch's Steam Steriliser*).

The articles to be sterilised are placed in a tin vessel provided with a lid, and the bottom of which is also grated, and are left in the steriliser for from half an hour to an hour (from the moment when an abundance of steam is given off), which suffices for complete sterilisation.

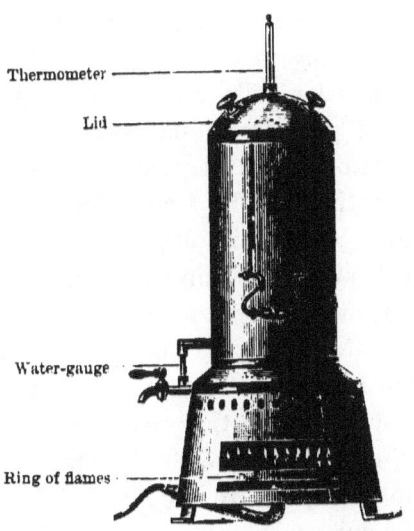

FIG. 10.—KOCH'S STEAM STERILISER.

For laboratories which are not supplied with gas, *Budenberg's steam generator* offers great advantages. In it the disinfecting cylinder communicates through a tube 4 cm. in diameter with a flat evaporating vessel, completely closed all round, in which water is heated to generate the steam. The disinfecting chamber is covered with a bell-shaped cylinder furnished with a thermometer, and serving the purpose of condensing the steam. The water so formed drips back into a dish, filled with distilled water before

C

heat is applied, which rests upon the upper wall of the evaporating vessel, and is connected with it by some small apertures.

Many substances, such as nutrient materials, are, on account of their albuminoid constituents, unable to stand the action of a temperature of 100° C. for any length of time without undergoing changes; gelatine, for example, loses the power of solidifying, which alone renders it applicable to bacteriological purposes. Hence it is advisable to expose all culture-media to a current of steam for not more than a quarter of an hour daily on three successive days. The heating on the first day kills all the micro-organisms present and most of the spores; but some of the latter still remain and develop by next day into micro-organisms, which perish on heating for the second time. Any that remain are destroyed on the third day.

Fractional sterilisation.—Certain substances, however, particularly the serum of blood, undergo so many changes even during a short sterilisation in the steam-current, as to be no longer suitable for use as culture-media, and in such cases recourse must be had to the process of *discontinuous* or *fractional sterilisation*, introduced by Tyndall. This consists of heating to a temperature of 54° to 56° C., for three or four hours daily during one week, in a chest with double walls between which there is a layer of water, the temperature being kept at a constant height by means of a thermo-regulator; or, according to Heim's method, the test-glasses can be placed in the warm water of a bath, the temperature of which is kept continually at the height mentioned above.

Sterilisation by steam under pressure.—High-pressure steam, applied by means of autoclaves, acts with greater rapidity than ordinary steam at 100° C.

Chemical disinfectants.—Besides high temperatures,

various *chemical* substances are employed for the purpose of sterilisation. Those possessing the greatest germicide power of all are *carbolic acid* in strong solutions, *corrosive sublimate* in 1 in 1,000 solutions, and *quicklime*; but next to these *chlorine, iodine,* and *bromine waters,* 1 per cent. solutions of *osmic acid,* 1 per cent. solutions of *potassium permanganate, oil of turpentine, iron perchloride,* &c., have a more or less energetic disinfectant action. Of the above disinfectants, corrosive sublimate in 1 in 1,000 solution is at present most in use.

Heider states that the efficacy of a large number of disinfectants is very markedly increased by moderately raising the temperature.

Chloroform is recommended by Kirchner as an excellent disinfectant, having the advantage of great activity combined with a low boiling-point, so that it can be driven off with certainty from other fluids by heating after sterilisation is complete. It is particularly suitable for sterilising blood serum, which cannot be exposed to a high temperature, and which, therefore, as Globig has shown, it is impossible to free by the method of discontinuous sterilisation from the germs of such micro-organisms as do not grow below 50° C., and are capable of withstanding a temperature of 70° C. To sterilise by this method the fluids under treatment are shaken up with excess of chloroform, and allowed to stand for some days, after which they are freed from the chloroform before use by heating for an hour at 62° C., the boiling-point of chloroform being 61·2°.

In bacteriological work it is often necessary to combine several modes of sterilisation in order to secure complete destruction of germs, but the method chosen varies continually according to the bodies to be so treated. For the sterilisation of instruments, boiling for five minutes in water is sufficient, according to Davidsohn; and plates, at all events, may be sterilised over a gas or spirit flame after being cleansed with alcohol and corrosive sublimate, after which they are laid, with the heated side uppermost, on a sheet of clean paper and merely protected with a glass cover which has been likewise cleaned by means of alcohol and corrosive sublimate, or even with a cleaned soup-plate.

Cleanliness of all objects coming in contact with the bacteria is of particular importance; and therefore it follows that in the practical application of bacteriology to surgery there is need of the utmost care in the cleansing of hands, instruments, and dressings, in order to render an aseptic procedure possible. For this purpose a thorough brushing of the hands (which have first been carefully cleansed with soap), followed by rinsing with alcohol and ether and washing in a $\frac{1}{10}$th per cent. solution of corrosive sublimate, is absolutely necessary.

Apparatus and Reagents

A **microscope** provided with Abbe's illuminating apparatus, ordinary objectives of various powers, and an oil-immersion lens.

The **steam steriliser** described above, with the corresponding gas-burner and a thermometer.

Incubator.—The warm chamber or incubator consists of a quadrangular chest of stout sheet metal with double walls, the space between which is filled with water and has two apertures, one for a thermometer dipping into the water, while into the other a *thermo-regulator* is inserted. The chest is closed above by a suitable lid, and the whole apparatus is covered with felt, with the exception only of the lower surface, to which heat is applied. The interior space may be subdivided by partitions.[1] A glass gauge,

---

[1] [The incubator chiefly used in this country differs slightly from that described in the text, as it opens at the side instead of above, and is closed

fixed to the outside, indicates the height of the water. The heating of the apparatus is carried on by small gas-flames (*micro-burners*), protected from draughts by cylinders of mica. The incubator serves the purpose of keeping cultures of micro-organisms at a fixed uniform temperature in cases where they will not grow at higher or lower degrees of heat (fig. 11).

In order to secure an even temperature a **thermo-regulator** is employed, which (with the very slightest varia-

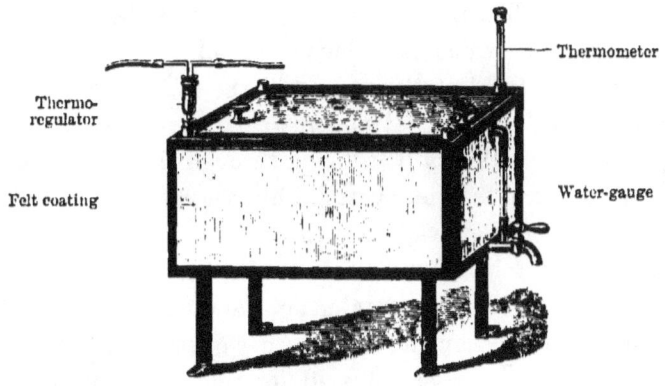

Fig. 11.—INCUBATOR.

tions) maintains the thermometer at a constant temperature.[1] It has the function of increasing the flame when the temperature falls, and diminishing it when the temperature rises by regulating the supply of gas. For this purpose Bunsen took

by means of double doors, one or both of which is made of glass, in order that the cultures, &c., in the interior may be observed without the loss of heat which must necessarily follow the opening of the apparatus. The two instruments are, however, identical in principle and in all other essential details.]—TR.

[1] [The terms 'room temperature,' 'ordinary temperature,' &c., which will be frequently met with in the following pages, denote a temperature of about 20° C., while by 'incubation temperature' is meant one of about the heat of the human body, i.e. 37° C. (German, *Zimmertemperatur* and *Brut-temperatur* respectively). The Incubator is usually kept at the latter temperature.]—TR.

advantage of the rise and fall of mercury, and, in case the opening for the passage of the gas should become completely closed, he devised a safety aperture which allows just so much gas to pass through as keeps the flame from being extinguished. Various thermo-regulators have been constructed on this principle.

Schenk's thermo-regulator.—A regulator in use for the incubator in the author's Institute is constructed as follows (it can be procured from Siebert, 19 Alserstrasse, Vienna): A piece of glass tubing is sealed into a vessel of glass shaped like a test-tube, in such a way that one end of the former, which is widened out, adheres air-tight to the sides of the latter vessel, while the other end reaches nearly to the bottom. If now mercury be poured into this apparatus a portion of it will sink through the narrow tube to the bottom of the wider vessel, while the rest fills the small tube and extends above it to such a height as to allow of the test-tube being closed with a cork which is perforated with one aperture. The air contained in the apparatus renders it more sensitive by causing the mercury to rise and fall more rapidly with the alterations in its volume produced by variations of temperature. Into the cork is fitted a second glass tube, which is expanded in its upper half, this expanded part being closed by a cork perforated twice and traversed by two glass tubes bent at right angles. One of these, which extends down as far as the constriction of the upper vessel, is ground away obliquely at the end, and has a minute safety aperture in one side; the other is simply a bent tube reaching to the lower surface of the cork only.

In actual use the apparatus is interposed between the supply-pipe of the gas and the burner by attaching to each angular tube a piece of india-rubber pipe, on the one side from the burner, on the other from the gas-tap. The test-

SCHENK'S THERMO-REGULATOR

tube-shaped vessel, supplied with a suitable amount of mercury, is placed in the water surrounding the incubator. On warming the water the column of mercury ascends so long as the gas flows from one angular tube to the other; when, however, the temperature becomes so high that the mercury reaches the longer angle tube which is obliquely ground, and closes the end, then the limit of temperature is fixed at which the incubator can be kept constantly.

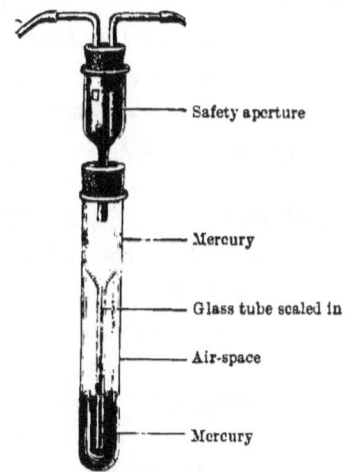

FIG. 12.—SCHENK'S THERMO-REGULATOR.

Now, as soon as this opening is covered with mercury, the passage of gas would necessarily cease and the flame of the burner be extinguished, did not enough gas escape through the lateral aperture to keep it burning. Accordingly, the filling of the regulator with mercury must be done in such a way that this limit is reached at the incubating temperature. By pushing in or withdrawing the glass vessel above the test-tube the temperature can be raised or lowered a few degrees according to need.

If the temperature of the water in which that part of the apparatus containing the mercury is placed rises, then the opening for the passage of the gas must become continually smaller, and this is followed by cooling of the water and sinking of the mercury, so that the aperture transmitting the gas again enlarges, and therewith the flame increases and the temperature ascends, but cannot pass beyond the limit for which the regulator is set. With good management the variations are only very small.

Meyer's thermo-regulator. — Another thermo-regulator, constructed by Victor Meyer, which is extensively used and can be highly recommended on account of its sensitiveness, consists also of a glass vessel like a test-tube, and which can be closed with a rubber cork. It is furnished with a small side-tube in the upper part, and is divided into two sections by a capillary funnel of glass, the end of which is just above the bottom. The lower division is filled with mercury, the surface of which is only some three cm. distant from the edge of the funnel, and the interspace thus left is occupied by a mixture of alcohol and ether. In the upper part is fitted a glass tube, cut off obliquely below, and passing through the rubber cork; it ends a little above the capillary funnel, and is pierced in one side above the lower opening with a hole the size of a pin's head.

In order to graduate the regulator it is immersed in a water-bath, the temperature of which is controlled by an accurate thermometer; even a slight increase of heat volatilises the ether and drives the mercury up, so that it comes to stand above the capillary funnel. If now the water has reached the temperature fixed on, the obliquely-cut glass tube is so far introduced into the mercury that the lower opening is quite covered and only the safety aperture at the side remains pervious. The regulator is so connected with the flame under the incubator that this receives only such a quantity of gas as can traverse the regulator. When the water attains too high a temperature, the ether vaporises and causes the mercury to rise, so as more and more to close the obliquely-cut tube until only the lateral hole remains open, and the discharge of gas is reduced to the minimum amount. If the temperature falls, the mixture of alcohol and ether again contracts, the mercury sinks, the supply of gas increases, and the temperature of the water rises once more.

Gärtner's thermo-regulator.—Gärtner has constructed a thermo-regulator which, instead of a safety-aperture, pos-

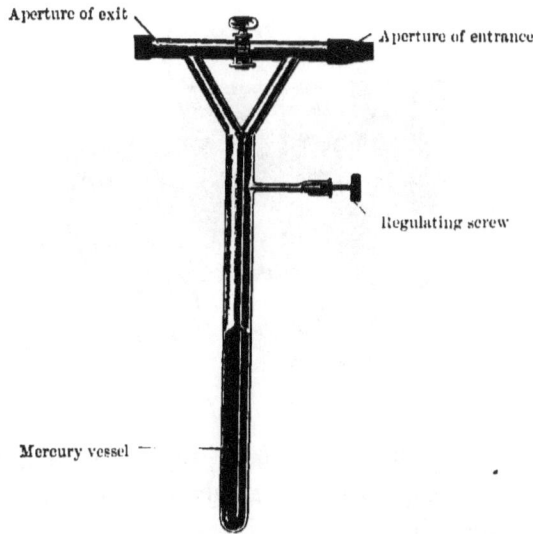

Fig. 13.—Altmann's Thermo-regulator.

sesses a by-road for the gas, consisting of a rubber gas-tube, compressible by means of a screw, to prevent the minimum flame, which must still burn even when the regulator proper has completely shut off the gas-supply, from proving too large if the pressure of gas increases.

Altmann's thermo-regulator.—The thermo-regulator of Altmann is constructed on Gärtner's principle, being provided with a horizontal tube which can be closed by a tap.

On each side of this tap a lateral tube proceeds in an oblique direction from the horizontal one, to unite at an acute angle with a vertical tube, which contains a vessel drawn out to a capillary termination, and filled with mercury in such a way that the convex surface of the metal begins at once to close the lumen of the angle formed by the union of the oblique tubes. If the lower end of the regulator is placed in too warm a medium the mercury expands in the capillary, and so permits the gas to pass only through the horizontal tube, where the supply can be still

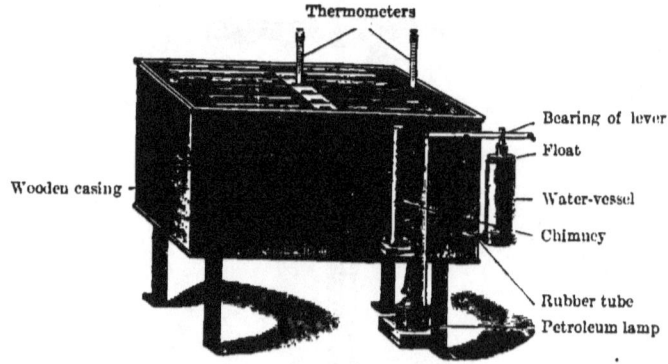

Fig. 14.—Baumeyer's Petroleum Incubator.

further reduced by means of the tap. The height of the mercury in the perpendicular tube can be regulated by a screw at the side.

Petroleum incubator.—If the laboratory is not provided with a gas-supply, incubators heated by petroleum can be employed, which are also provided with contrivances arranged for regulating the temperature in the chamber. Such an incubator is made of wood, and contains within it a chest of sheet metal, to the bottom of which is fitted a heating canal furnished with a chimney which runs perpendicularly up the outside wall (fig. 14). The flame of a petroleum lamp plays into the heating canal and warms the water circulating in six rubber pipes connected in a water-tight manner with the metal chest. A vertical tube is so attached to the outside that the surface of the water in it ascends as the temperature rises, and it contains a float furnished with a lever which presses upon a bar connected with the lamp and made to regulate the flame of the petroleum. When the surface of the water sinks the bar is lifted and the flame increased; when the water ascends the flame is diminished (Baumeyer's apparatus).

Water-baths and **sand-baths**, and such contrivances for warming and boiling, with suitable stands, **gas-burners**, an **ice-tank** for refrigerating, **flasks**, **test-glasses**, and **funnels and dishes** of the most various kinds are also included in the equipment necessary for a bacteriological laboratory.

Hot-water filter.—A kind of funnel known as the hot-water filtering funnel is frequently in use when it is

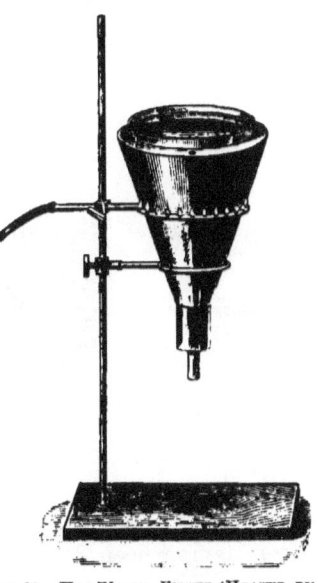

FIG. 15.—HOT-WATER FILTER (HEATED BY A RING BURNER).

necessary to filter in the hot state substances which become solid at ordinary temperatures. Such a funnel is made of copper, brass, or sheet-iron, with double walls, and fitted with an appendage at the side which is warmed by means of a flame; but those hot-water funnels in which the warming is managed by a ring burner surrounding the lower part of the external surface are also very efficient. By filling the space between the walls through an opening above with water, which is then heated, masses of gelatine and agar can be filtered while warm through a suitable glass funnel inserted into the hot-water filter (fig. 15).

Apparatus for plate-cultivations.—For the purpose of spreading nutrient gelatine upon plates a levelling apparatus is in use which must be so arranged that the plate lies horizontally, in order to prevent the gelatine from easily running off it when poured out. The apparatus consists of a levelling-stand in the shape of a wooden triangle with feet formed by levelling-screws, upon which rests a rather

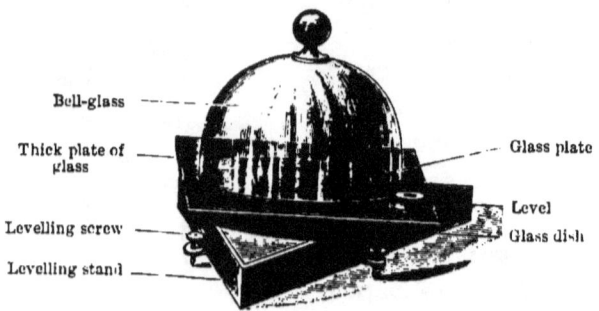

FIG. 16.—LEVELLING APPARATUS FOR MAKING PLATE-CULTIVATIONS.

large glass dish filled with water and pieces of ice and covered with a thick glass plate or a sheet of iron. The latter having been brought into a horizontal position with the help of a spirit-level, glass plates to receive the gelatine can be laid upon it and protected with a bell-glass (fig. 16).

Moist chambers are employed for the further carrying out of the cultures; they are made of glass, and have a diameter of about 24 cm. and a height of 6 to 7 cm. (see p. 56).

Instead of the ordinary **glass plates**, which are the size of a photographic quarter-plate, round glass dishes are also used. **Petri's capsules** consist of flat double dishes of glass, of which the lower has a diameter of 10 cm.

REAGENTS USED IN BACTERIOLOGICAL RESEARCH

Soyka's plates are similar to Petri's capsules, but differ from them in having eight to ten depressions ground in the lower plate, which resemble the 'wells' in hollowed slides.

In addition to the above, all articles employed in microscopic investigation are required. The slides used for examining micro-organisms in the '*hanging drop*' have a well ground in the centre (fig. 17). This is covered with a cover-glass the lower surface of which has been prepared with the micro-organism.

Crates of galvanised wire are used for holding glass utensils, especially test-tubes, while being sterilised (fig. 18).

Reagents.—It is scarcely possible to give a complete list

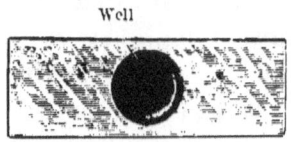

FIG. 17.—HOLLOW SLIDE.

FIG. 18.—WIRE CRATE.

of all the reagents used in bacteriological research, since recourse must be had as much as possible, according to the nature of the particular investigation, to the province of the auxiliary sciences, including, of course, chemistry. Speaking generally, the reagents used are acids, salts, disinfecting fluids, various oils, colouring matters, and other drugs.

Of *acids*, those most used are sulphuric, nitric, chromic, acetic, and oxalic, and less frequently also dilute osmic acid.

Of *alkalis and salts*, solutions of caustic potash and soda and lime water are employed, also the bicarbonates of potassium and sodium, sodium chloride, potassium iodide, iron perchloride, ammonium carbonate, and potash-ammonia alum.

Iodine is used both solid and in solution.

Chloroform is an important disinfectant, particularly for sterilising blood-serum.

Corrosive sublimate in solutions of 1 to 1,000 and *carbolic acid* are necessary reagents for the laboratory table.

The *oils* employed are aniline, cedar, and those of bergamot and cloves, used partly for clearing the microscopic preparations, partly as solvents.

For imbedding, a harder and a softer variety of *paraffine*, and *celloidine*, are used.

Canada balsam, and less frequently *glycerine*, are employed in the preparation of permanent microscopic objects. The latter finds, however, its most extended application in preparing nutrient materials.

Gelatine, agar-agar, blood-serum, albumen from the eggs of hens and of insessorial birds, *potatoes, starch, paste, milk, rice,* and *bread* are all used for making nutrient substances. Their application will be gone into in detail in treating of culture-media.

Stains.—The following is a list of the colouring matters which are indispensable in making bacteriological preparations:—*Fuchsine, methyl blue, gentian violet, Bismarck brown, methyl violet, malachite green, eosine, safranine,* and *dahlia* are aniline colours which are sufficient for nearly all investigations. Besides these, however, *carmine, picrocarmine, picric acid, hæmatoxyline,* and *Magdala red* are required for preparations of tissues; and for certain methods of staining still other dyes are used, such as *extract of logwood*,[1] &c.

Distilled water, alcohol, ether, xylol, and *oil of turpentine* complete the equipment of reagents.

Other utensils.—*Platinum wires* with loops sealed into

[1] [Practically identical with *Extractum Hæmatoxyli*, B. P.]—Tr.

MISCELLANEOUS APPARATUS

glass rods—best done over a Bunsen burner or with the aid of a gas blow-pipe.

Of *instruments*, scalpels, scissors, forceps, different kinds of needles, hooks, inoculation needles, and hypodermic syringes, or better still Koch's syringe (fig. 19), are required; and, in the preparation of nutrient media and their use for cultivation, flasks, test-tubes, dishes, plates, pipettes, blocks or benches of glass, potato-knives, &c., are employed.

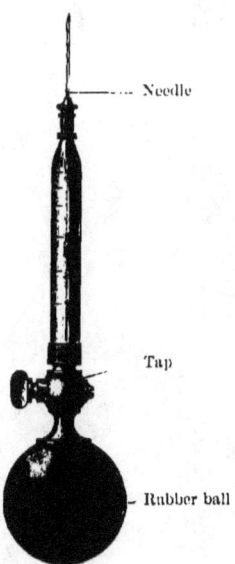

FIG. 19.—KOCH'S INJECTING SYRINGE.

With the instruments now specified, a laboratory is in a position to begin bacteriological work, and it need only be mentioned that the articles necessary for all scientific work must also be at hand, as, for example, working benches, cupboards for reagents, test-tube stands, corks, meat-presses, scales, different kinds of glass and metal vessels, bibulous paper, &c.

Centrifugal machine.—In order to examine fluids which are poor in corpuscular elements, Stenbeck has introduced a centrifugal machine, or *centrifuge*, to be driven by hand. This contrivance carries a metal frame or a disc with several apertures in which are fixed metal cases for the reception of small glass tubes. The fluid to be examined is poured

FIG. 20.—STENBECK'S CENTRIFUGE (after Jaksch).

into the small tubes, which are provided at their lower end with a little reservoir communicating with them through a conical constriction, and in which the precipitate gathers when thrown to the bottom. This centrifugal machine has been several times modified by Von Jaksch (fig. 20).

CENTRIFUGAL MACHINES

Gärtner's centrifuge consists of a case of sheet brass with a movable cover. The bottom is shaped like the surface of a very flat cone, and carries clamps for small test-tubes, which are laid in with their mouths towards the centre, and are so far filled with the fluid to be treated that none flows out when the tubes are placed in the slanting position. Six to eight samples can be centrifuged at the same time. As soon as the glasses are in position, the cover is lowered and fastened down by means of a bayonet-catch. The centrifugal machine has for its axis a spindle

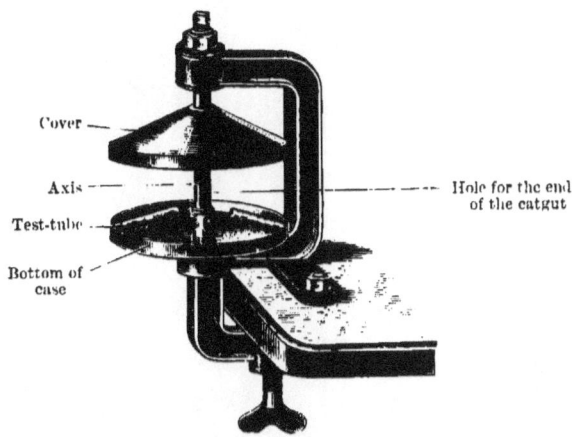

FIG. 21.—GÄRTNER'S CENTRIFUGE.

revolving with very little friction in sockets in a cast-iron frame, which serves to fasten the entire apparatus to a table or a window-sill. A hole is perforated in the lower part of the spindle, into which the end of a catgut string is introduced, the string itself being wound round the shaft. By pulling away the string the apparatus is put in rotation after the manner of a child's top, the number of revolutions at starting reaching over 3,000 in the minute, and the motion keeps up with gradual slackening for ten to fifteen minutes (fig. 21). For decanting the supernatant liquid

Gärtner uses a little contrivance consisting of a cork with two glass tubes which is inserted into the mouth of the test-tubes, and converts them into miniature wash-bottles. The fluid is forced out by blowing and the sediment remains behind on the bottom.

Less recently Csokor constructed a large centrifugal machine marked by its fixed and perfectly even rate of rotation. It is driven by water-power, and makes over 3,000 revolutions per minute.

CHAPTER III

NUTRIENT MATERIALS AND METHODS OF CULTIVATION

Nutrient media.—In order to observe the growth of micro-organisms, it is absolutely necessary to provide a number of nutrient materials in which the individual microbes may multiply into larger masses, so that their peculiarities can be more thoroughly made out. Some of these culture media are so prepared as to approximate more or less closely to the natural soil of the micro-organisms, and others in such a manner as to render them suitable for use as general media on which the most widely differing varieties may be cultivated. They are divided generally into *liquid* and *solid* media.

Liquid nutrient media.—Fluid media fall rather into the background in use compared with the solid, since the conditions of growth and characteristic peculiarities in the shape of the colonies come out less strongly on them than on the latter.

They are employed either in sterilised test tubes closed with a plug of cottonwool, or in little flasks, of which those of Erlenmeyer are particularly useful (fig. 22). The media, especially *bouillon* or broth, after being distributed into such smaller vessels, must be carefully heated in the current of steam for 15 minutes daily on three to five successive days, in order to sterilise them. The

FIG. 22.
ERLENMEYER'S FLASK.

inoculation and further cultivation of the micro-organisms depends on the particular kind under observation.

Preparation of meat bouillon.—The following recipe gives Löffler's method of preparing a liquid medium (*broth or bouillon*) which has come into general use :—A half-kilogram weight of meat freed from fat is chopped fine in a mincing machine. (Such meat cannot, strictly speaking, be termed free from fat, because only the masses of fat are cleared away, whereas that which exists in the substance of all meat is not removed.) A litre of ordinary water is poured over the meat and the whole is allowed to stand in a cool place for twenty-four hours. In this way the albuminoid bodies and other substances soluble in water are dissolved out from the meat, the result being an aqueous extract, coloured in most cases with hæmoglobin, and which is separated from the residue by squeezing it through a cloth. About a litre of fluid is thus obtained, which must now be freed from albuminoid bodies by heating in a water-bath or in the steam-steriliser. The heating must be kept up until a sample, when filtered and boiled, no longer shows any turbidity, which usually takes about half an hour. The fluid is then filtered, and to the filtrate are added one per cent. of dry colourless peptone and 0·5 per cent. common salt, which is equivalent to 10 grams peptone and 5 grams salt to the litre of water. After this the solution is boiled, and, as it has a feebly acid reaction, is neutralised with a saturated solution of sodium carbonate, until it causes in litmus paper a slight blue coloration, which afterwards passes into a faint red. It is not essential to add water to make up what has been lost through evaporation, but it will do no harm to do so if desired. The broth thus prepared is boiled once more and filtered after boiling; it should not become turbid, either during sterilisation or on standing. The filtrate must be clear, pale yellow, and of neutral or feebly alkaline reaction. Turbidities are either caused by the reaction being strongly alkaline, and are in that case removed when this is corrected, or are due to a finely flocculent precipitate of albuminates, which are cleared away by adding the white of a hen's egg and boiling for a quarter of an hour [with subsequent filtration].

Preparation of meat extract bouillon.—A second mode of making bouillon consists in the combination of meat-extract and sugar to obtain a liquid culture medium. The following is Hueppe's process:—To a litre of water are added $\frac{1}{2}$ per cent (5 grams) of extract of meat and 3 per cent. (30 grams) of dry peptone, or instead of extract of meat and peptone, 2 to 3 per cent. (20–30 grams) of peptone of meat. A further addition of five grams of grape or raw sugar is then made, and the liquid boiled, carefully neutralised with solution of sodium carbonate, and subsequently sterilised. Admixture of glycerine with the bouillon is also advantageous, and tubercle bacilli grow excellently on the medium thus obtained.

Solutions of white of egg.—Solutions of white of egg are well suited to form fluid culture-media after they have been completely freed from germs by the discontinuous method of sterilisation. The white of plovers' eggs lends itself well to this purpose, being clear and transparent, and capable of dilution with water, and of being filtered; it admits also of the addition of dextrine, sugar, or other ngredients according to need. The albumen when sterilised affords for a considerable time a suitable medium for cultivations in test tubes or on glass plates placed in the moist chamber.

Solid nutrient media.—Owing to the introduction of the use of solid culture media in bacteriological research, a series of micro-organisms have been more thoroughly examined, and it has thus been possible to observe that certain characteristics in the growth and multiplication of the elements in this way are more distinctly brought out and that the individual forms are thereby specially distinguished.

Preparation of peptone bouillon gelatine.—The Koch-Löffler peptone bouillon gelatine is that most in use, and is prepared in the following manner :—500 grms. of meat freed from fat are minced up fine, mixed with a litre of water, and allowed to stand for twenty-four hours, after which the mass of meat is squeezed out, the filtrate boiled in a water-bath for three quarters of an hour until all albuminoid bodies are precipitated, and then filtered again. Another method is to place the meat, over which a litre of water has been poured, at once upon the fire, between the flame of which and the vessel a plate of asbestos must be interposed. The broth is made to boil for several hours and then let cool, in order to separate the fat, after which it is filtered and sufficient water poured in to replace that lost by evaporation. 100 grms. gelatine, 10 grms. colourless peptone, and 5 grms. common salt are next added to the filtrate, and this mixture is allowed to stand for some time and then heated in the water-bath until all the gelatine is dissolved. In order that the gelatine may be colourless, the solution must not become concentrated while being heated in the water-bath, but from the very commencement as much water must be added, from time to time, as it appears to have lost by evaporation. The reaction of gelatine is always acid, and this is also the case with broth, so that it must be neutralised with a concentrated solution of sodium carbonate or with solution of caustic soda.

The entire mass is next filtered through a creased filter paper in the hot-water funnel.[1] The creased paper must be moistened with warm water before filtering, as otherwise

---

[1] [The paper is not folded in the manner usually adopted, but creased in folds radiating from the centre, somewhat like a circular fan; the piece

the pores would be clogged by solidification of the gelatine. To avoid using the hot-water funnel, Kirchner suggests allowing the gelatine to cool slowly in the steam steriliser, after turning out the flame; it is then after a few hours perfectly clear and can be filtered with facility. Instead of the folded paper, a thin layer of cotton- or glass-wool may be used for filtering.

If a sample of the filtrate is taken in a test-tube and heated until it boils, it must remain clear and should also not become cloudy while cooling. Any turbidity, if such occurs, may possibly be due to the gelatine having been rendered too strongly alkaline in neutralising; as in heating such a solution the carbonic acid is driven off, and then compounds are thrown down which cause the turbidity. It need hardly be said that care must be taken in such cases to neutralise exactly in order to get an efficient gelatine. That perhaps other faults, such as inferior paper, dirty vessels, &c., may be to blame, is also evident, and such must be avoided in the preparation of nutrient media. Cloudiness is most easily dealt with by adding the white of a hen's egg to the lukewarm gelatine while it is still fluid, and shaking so as to divide it finely, after which the solution is again boiled and filtered in the hot-water filter. Indeed it is the rule to add the white of an egg to nutrient gelatine immediately after neutralising, so as to ensure the avoidance of all faults of turbidity.

The gelatine when ready should be clear and of an amber-yellow colour, and should not become cloudy on heating. Carefully cleaned test-tubes are filled with about 10 cubic centimetres each and plugged with cotton wool, or Schill's double test-glasses may be used (see p. 16). The

should be about eighteen inches square, and folding is begun by doubling it down the centre. The creased paper is finally gathered up, inserted into the funnel, and the superflous part cut off.]—Tr.

test-tubes are sterilised before filling, by placing them one over the other in a wire crate lying on its side, in which they are introduced into the hot-air steriliser and exposed for an hour to a temperature of 100° C. [The cotton-wool plugs should be inserted before sterilising.]

The gelatine is introduced into the test-tubes with the aid of a pipette, care being taken that it does not soil the edge of the tube, and least of all comes in contact with that part of the inner surface which supports the cotton plug. In this manipulation the plug must be seized on the dorsal surface of the hand between two fingers, and extracted from the tube with a twisting motion ; the pipette is then filled and closed with the forefinger, which is only raised when the gelatine is to be allowed to run out into the tube. After this procedure has been repeated a few times, each worker in his own way acquires such a degree of expertness, that the greatest possible celerity is attained in filling the tubes with gelatine and quickly reclosing them.

Instead of the pipette, the use of which always demands a certain amount of skill, small glass funnels may aptly be employed, or a glass tube capable of being closed by a tap may be attached to the funnel through which the gelatine is filtered and introduced into the test-tube to be filled.

It is particularly to be observed that the gelatine must not be kept continuously at a high temperature, lest it should lose its power of solidifying when cold. It must, therefore, be heated in the steam apparatus for fifteen minutes on several—about three to five—days in succession, in order that the culture-medium preserved in the test-tubes may be completely sterile, and capable of being stored for future use in all bacteriological experiments.

Preparation of meat extract peptone gelatine.—Hueppe's meat-extract peptone gelatine is a 10 per cent. solution of gelatine, to which 5 grm. extract of meat, 5 grm. grape-sugar, and 30 grm. peptone have been added. As soon as the gelatine is dissolved in water the other ingredients are mixed in and the whole boiled, after which the solution is filtered off by means of the hot-water funnel and neutralised. Should clouding by any chance occur, recourse must be had to those measures described in speaking of the preparation of peptone bouillon gelatine. When this has been sterilised—which must be done with particular care, as the extract of meat contains many germs—a culture-medium is obtained which can in most cases be used exactly like the preceding. The finished gelatine is stained a brownish colour, owing to pigments derived from the meat extract.

Both these modes of preparation yield gelatines which are most extensively used in all bacteriological researches. Attempts are now made to alter the culture-media by adding various substances to the gelatine, such as *grape-sugar* (up to 2 per cent.) and *dextrine* or *glycerine* (4 to 6 per cent.). By means of these different admixtures, the nutritive value of the gelatine for certain micro-organisms is said to be increased.

In summer the gelatine has a tendency to liquefy, and its strength must accordingly be increased from 10 to 15 per cent.; while for cultivating anaerobes a $7\frac{1}{2}$ per cent. gelatine is required.

Additions to nutrient gelatine.—A modification deserving of special notice is *litmus gelatine*, which is prepared by mixing a tolerably concentrated solution of blue litmus with the gelatine, thus obtaining the substance to which this name is given. Its importance lies in the fact that the acids or alkalies formed by micro-organisms in their growth can thus be qualitatively demonstrated.

It is advisable to add the most widely different substances to the gelatine, so as to meet individual requirements in cultivating micro-organisms, and accordingly the greatest multiplicity of admixtures have been recommended by approved investigators. For example, Koch uses mixtures of gelatine with *blood-serum, aqueous humour, infusion of hay and of wheat, decoction of horses' dung,* and *decoction of plums.*

Miquel uses, instead of meat bouillon, a solution of 40 parts peptone, 10 parts common salt, and 1 part carbonate of potash in 1,000 parts of water, to which the further addition of 4 parts of gelatine can be made.

Holtz has devised a *potato-gelatine,* for use in growing typhoid bacilli. The potatoes are grated and squeezed through a straining-cloth, the liquid which flows away is allowed to stand for twenty-four hours, and is then boiled with 10 per cent. of gelatine.

Preparation of urine gelatine.—A very cheap and easily-prepared gelatine is the urine gelatine recommended by Heller. Urine is caught in sterilised vessels, and its specific gravity having been brought to 1010 by dilution with sterilised water, it is rendered feebly alkaline with soda solution and filtered. After 1 per cent. of peptone, $\frac{1}{2}$ per cent. of common salt, and 5 to 10 per cent. of gelatine have been added, it is next boiled and filtered, and the fluid so obtained is poured into test-glasses and sterilised, a single sterilisation being said to be sufficient. This process can be modified by filtering the urine through animal charcoal before diluting it with water, in order to remove part of the urinary colouring matter.

Preparation of nutrient agar.—Agar-agar is a vegetable jelly procured from different algæ growing in the East Indies and Japan, and was introduced into bacteriology by Hesse because of its distinctive property of remaining in the solid state at 40° C., and only melting completely at

90° C. Hence this jelly is well adapted for use as a culture-medium for those micro-organisms which must be grown at the higher temperatures in the incubating chamber. Agar-agar appears commercially in the form of transparent strips, or four-cornered pieces, or as a white powder, and swells up in water.

Preparation of peptone bouillon agar.—To make nutrient agar (*peptone broth agar*), 500 grm. of meat free from fat are taken, minced up, and mixed with a litre of water, and after standing for twenty-four hours in a cool place the liquid is filtered through a cloth and squeezed out from the mass of meat. The albuminoid bodies are precipitated by boiling the meat infusion, and are removed from the liquid by filtration, the result being ordinary nutrient bouillon. This is rendered feebly alkaline with sodium bicarbonate, and mixed with 10 grm. peptone, 5 grm. common salt, and 20 grm. agar-agar cut up small. The agar swells up in the broth, and is then boiled over a sand-bath, in the steam steriliser, or even over the naked flame, until only small flakes and slight turbidities are observed, the fluid lost by evaporation being then made up by the addition of water. It is seldom necessary to neutralise the fluid a second time with sodium bicarbonate, as the agar has of itself a neutral reaction, so that it only remains to filter the solution, though this is in some cases difficult to do successfully. The filtration is carried on through a double layer of filter-paper, by means of the hot-water funnel, or in the steam steriliser, and is a very slow process. After addition of white of egg it sometimes happens that when the agar mass has been boiled the small particles are gathered into lumps by the albumen, and the filtration is much easier in consequence.

Some recommend that the hot agar should be allowed to cool gradually in a tall cylindrical vessel placed in the steam steriliser, by which means the turbidities sink to the bottom, so that the solidfied mass of agar is perfectly clear in the upper part. This mass is removed from the glass by slightly warming it, the clear part is separated from the turbid by cutting it with a knife, and is then chopped up and re-melted.

To simplify the preparation of nutrient agar it is advisable to dissolve the finely-cut pieces of agar in boiling water over the open fire, a sheet of asbestos being interposed between the flame and the vessel; this takes from half to three-quarters of an hour, and the agar should not be mixed with the broth until solution is complete. In this case also it is better not to heat the liquid too long, as the medium becomes dark if allowed to boil away, but from the first to add as much more water from time to time as might be lost through evaporation. In order to secure a clear solution, the pieces of agar may be first of all laid in 2 per cent. hydrochloric or 5 per cent. acetic acid, which is afterwards washed away with water. By Richter's method the agar is dissolved in wine by two hours' maceration and subsequent boiling, and the solution is added to bouillon.

Tischutkin allows the required quantity of agar to swell for 15 minutes in a very dilute solution of acetic acid, washes it in pure water, and thereupon adds it to the broth, in which it dissolves after only 3 to 5 minutes' boiling. When it has been neutralised and cooled, the whites of two hen's eggs are poured in, and the mixture kept in the steam apparatus for half to three-quarters of an hour. The subsequent filtration occupies only a short time, even without the hot-water funnel.

The agar when ready is filled into sterilised test-tubes closed with cotton-wool, in such a way that about a third of the test-tube is occupied by the fluid. Care must next be taken that the nutrient mass be sterilised, which is done by heating the filled test-tubes in the steam steriliser for twenty minutes daily on three successive days.

The tubes when filled and sterilised are laid in a slanting position, for which purpose suitable contrivances of various kinds are employed, the result being that the surface of the agar after setting forms a very acute angle with the long axis of the tube. During solidification some water, the *water of condensation*, separates out and prevents the firm adherence of the agar to the vessel, so that the medium often turns if the test-tube is rotated, a phenomenon which does not disappear until the water of condensation has evaporated. Esmarch recommends the addition of some *gum arabic* to prevent slipping away from the surface of the glass.

The mass of agar is clear and transparent while liquid, but after solidifying is somewhat cloudy and opaque.

Nutrient agar is often prepared by adding 20 grm. agar, 5 grm. extract of meat, 5 grm. grape-sugar, and 30 grm. peptone to a litre of water. Sterilisation must be carried out with greater care than in the case of ordinary nutrient agar, and the prepared medium is of a brownish-yellow colour.

The most varied modifications of nutrient agar can be produced by the addition of solution of litmus, grape-sugar, and other substances soluble in water. That most commonly used is the *litmus agar* mass prepared by adding 40 ccm. solution of litmus to a litre of prepared agar. This medium is *par excellence* of service in the carrying on of researches on the formation of acids or alkalies during the growth of micro-organisms.

Modifications of gelatine and agar, &c.—*Glycerine agar* is often made use of, as several micro-organisms grow very readily on this medium, which consists of nutrient agar modified by the admixture of about 6 per cent. of pure neutral glycerine. It has lately been shown by Nocard and Roux that tubercle bacilli and the bacilli of glanders flourish very freely on a nutrient medium so made.

Kowalski prepares both nutrient gelatine and nutrient agar in the following way:—Instead of meat, a kilogram of calf's lung is used in the preparation of the broth. It is cut up, two litres of water poured over it, the fluid allowed to stand in a cool room for some time, and squeezed out after boiling. To the filtrate so obtained are added 25 grm. peptone, 90 grm. sugar, 18 grm. common salt, 9 grm. sodium phosphate, 9 grm. ammonium sulphate, and 25 grm. sodium sulphate, and as soon as these ingredients are all dissolved a further addition of 10 to 15 per cent. of gelatine or 2 per cent. of agar-agar is made, and the whole is boiled with continual stirring. After the mass has cooled, but before it has become viscous, the whites of five hen's eggs are mixed in and divided up in the fluid, and when it has been boiled once more until all the albumen is coagulated, 8 to 10 per cent. of glycerine is added. This culture-medium is, when clear, of a straw-yellow colour. It is distributed into test-tubes with a pipette, and sterilised in the steam apparatus for ten minutes on three days in succession. Kowalski has described very favourable results obtained with it.

The method given by Heller for the preparation of urine gelatine can also be used to produce *urine agar*; the process already given is followed, 1 to 2 per cent. of agar being added instead of the gelatine.

Gelatine as well as agar media may be prepared with *milk* or *caseine*, after the method of Marie Raskin, and peptone can be used as an addition, as also albuminate of soda or potash.

Miquel has described a nutrient jelly which he prepares from an Irish moss (*Carragheen, Fucus crispus*) by boiling 300 to 400 grms. of it with 10 litres of water, and filtering, the filtrate being then evaporated and dried at 40° to 45° C. By the addition of 1 per cent. of this extract to broth, a solid nutrient medium is obtained, which only begins to liquefy at 50° C.

Besides the modifications of nutrient broth, gelatine, and agar, just mentioned, numerous other alterations and additions have been proposed, the application of which is, however, very limited.

Blood serum.—The use of blood-serum as a nutrient medium was introduced into the practice of bacteriology by Koch. The blood from a punctured or incised wound is allowed to flow into a sterilised tall glass cylinder, which is then placed in an ice-tank, and allowed to stand undisturbed for forty-eight hours. In this way the serum, which should be of a yellow or pale red colour, separates out, and it is then poured with the aid of a sterilised pipette into sterilised test-tubes plugged with cotton-wool, so that there are about 10 ccm. of serum in each tube. The liquid serum is exposed to a temperature of 56° two hours daily for a week, and freed from germs by this fractional sterilisation. In order, however, to kill those micro-organisms also which grow at a higher temperature, it is next shaken with chloroform in excess, and allowed to stand for a few days, the chloroform being removed by heating before use. The test-tubes are then laid on a slanting surface and the serum made to set at a temperature of 70° C. When solid, it should have a jelly-like consistence and a yellowish colour, and should adhere in its whole extent to the test-tube as a transparent mass. Koch has devised a special apparatus for the inspissation of blood serum, in which the water is carefully heated for about half an hour to 68°–70° C. (fig. 23). It becomes opaque at higher temperatures. Before use it must be ascertained whether the serum is sterile, which is most readily done by covering the test-tubes with india-rubber caps and letting them stand for a few days in the incubator. Only those test-tubes should be used in which no germs of micro-organisms have visibly developed.

The serum of human blood is often employed, and can be obtained sometimes at operations and sometimes from placentæ.

FIG. 23.—APPARATUS FOR THE INSPISSATION OF BLOOD SERUM.

Modifications of serum.—Those fluids which are procured from hydroceles, ovarian cysts, or dropsical effusions, are very nearly akin to human blood-serum, and the process for manufacturing nutrient media from them is similar.

Löffler modified the blood-serum as follows :—Having freed an aqueous extract of meat from albuminoid bodies, he added 1 per cent. of peptone, 1 per cent. of grape-sugar, and 0·5 per cent. of common salt to it. This solution, which has an acid reaction, is neutralised with sodium bicarbonate, then sterilised in the steam apparatus, and, after cooling, mixed with liquid blood serum, in the proportion of one part broth to three parts serum. Test-tubes having been filled with the mixture and sterilised by the discontinuous method, the mass is inspissated at 70° C.

Admixture of 6 to 8 per cent. of glycerine is recommended, and Hüppe also makes *serum gelatine* by adding a concentrated gelatine solution, or *serum agar*, with a 2 per cent. solution of agar, in the proportion of two to three parts of either to one part of serum; the resulting nutrient media being freed from germs by fractional sterilisation. These have recently been a good deal used.

Eggs of birds.—It is well known that the white of hens' eggs, when mixed with a concentrated solution of potash and poured from one vessel to another, coagulates to a rather firm jelly, which is known as Lieberkuhn's *potash albuminate*. Taking advantage of this process, Tarchanoff and Kolesnikoff have accordingly employed as a nutrient medium an alkaline albuminate prepared in the following way:—Hens' eggs are laid without being denuded of their shells in a 5 to 10 per cent. solution of potash for about fourteen days. In this way the white becomes firm like gelatine, probably owing to a combination between the albumen and potash taking place through the pores of the calcareous shell and the membrane more slowly than is the case in preparing Lieberkuhn's potash albuminate. If this pale yellowish jelly-like mass of albumen is cut into fine slices, and the strong alkalinity got rid of by washing, a very useful nutrient medium is obtained. It need hardly be said that thorough attention must be paid to sterilisation, which can be done in the steam steriliser.

Plovers' egg albumen.—A convenient albuminous nutrient material, which commends itself on account of its great transparency and colourlessness, is the white from the eggs of insessorial birds up to the time when the embryo has developed its vascular area. This medium was introduced and has been frequently used in the author's Institute. We employ for this purpose the white of plovers' eggs, as these are easily bought in the spring time. When a plover's egg is opened after washing with corrosive sublimate solution, there is found round the vitelline membrane a condensed mass of albumen, outside of which the white is clearer and less dense. If this outer white is distributed into narrow test-tubes, and these laid slanting and subjected to a temperature sufficient to coagulate the albumen, a clear gelatinous transparent mass is obtained, which can be used for the most widely-differing cultures: for example, even gonococci will grow upon it, as Von Schrötter and F. Winkler have shown in the author's Institute, and pigment-forming micro-organisms develop particularly well upon this medium.

Admixtures of various other substances—grape-sugar, dextrine, paste, and in fact all bodies soluble in water, and which have not an acid reaction—can be made with it, so as to modify it in various ways. The medium can also be prepared by diluting the concentrated albuminous mass with water but in that case the white of egg must first be filtered.

Although investigations show that no micro-organisms can be detected in the normal unfecundated plover's egg, it is nevertheless advisable to subject the test-tubes when filled to a fractional sterilisation before use.

Plover's egg albumen is also applicable to plate cultures, but after careful sterilisation it must be dried on a sterile glass plate over sulphuric acid in the receiver (also sterilised) of an air-pump, and then kept in the moist chamber after inoculation. The preparation of plates may, however, be facilitated by mixing the albumen with gelatine or agar.

Hens' eggs.—According to Hüppe and Heim, hen's eggs may themselves be used with advantage as nutrient media.

Fresh eggs are cleansed with soda solution, washed, and laid in corrosive sublimate, which must be removed with ammonium sulphide, spirit, and ether before they are inoculated. The latter is done by piercing the apex with a needle which has been sterilised at a glowing heat, and introducing the seed material into the interior by means of a glass capillary tube, from which it is carefully blown out as the tube is withdrawn. Closure is then effected with sterilised cotton-wool or collodion. This method is particularly well adapted for the cultivation of anaerobes.

Potatoes.—An excellent culture material for the different micro-organisms, and one, moreover, which secures their development in a quite characteristic way, is the potato. This medium is prepared as follows, according to Koch's method:—The potatoes are carefully cleaned in water with a brush, and after being freed from dirt are laid for half an hour in a 1 per 1,000 solution of corrosive sublimate to which some $\frac{1}{2}$ per cent. hydrochloric acid has been added, and finally washed with water to remove the adherent sublimate. With the aid of a sterilised knife single specks of dirt and 'eyes' are removed from the surface of the potatoes, which are next heated for about an hour in the steam apparatus and then cut into halves with a sterilised knife, when the free cut surface of each of the halves can be utilised for the culture of micro-organisms. If it is desired to ascertain whether the potatoes are perfectly sterilised, several of those treated as above can be left in moist chambers and watched to see whether micro-organisms develope, in which case the process of sterilising in the steam apparatus must be repeated.

If the potatoes after being washed in water, disinfected in corrosive sublimate, and carefully cleaned on the surface, are cut into several rather thick discs which fit into small sterilised boxes of glass, each individual disc can be used as a medium upon which to cultivate (*Esmarch's potato discs*); but they must be separately sterilised on from three to five successive days in the steam steriliser. They may also be kept stored in several glass dishes standing one above the other, and which mutually cover and close each other.

In order to use potatoes in a transparent form, thin slices can, after Wood's method, be cut from very white potatoes, and firmly pressed upon sterilised slips of glass, which are introduced into test-glasses and then sterilised.

Instead of discs, cylinders may be punched with the aid of a cork-borer out of cleaned and peeled potatoes. Each cylinder can be split into two halves in its long axis, and each half placed in a test-tube closed with cotton-wool. After these test-tubes have been sterilised in the steam apparatus for three days in succession, the surface of the piece of potato is inoculated. It is advisable to support the potato cylinders on cotton-wool or small glass tubes, which can absorb the water of condensation.

Potatoes are often pounded up after being peeled and boiled, and are then pressed into little Erlenmeyer's flasks (*potato pap*); and in this way a useful culture medium is obtained after proper sterilisation. Eisenberg has modified the process by using, instead of flasks, small boxes capable of being closed with a glass lid, and by sealing these with paraffin for permanent cultures.

Nutrient materials made from potatoes possess a slightly acid reaction, so that the surface of the potato must be rendered alkaline with sodium bicarbonate solution for certain micro-organisms, whose growth demands alkalinity.

Rice, bread, and wafers.—*Milk rice* is prepared by mixing skim-milk with two and a half times its quantity of powdered rice. This mixture is boiled until a thick pap is obtained, which is slightly cooled and introduced into a cork-borer so that no air-spaces remain. The pap is then pushed out with a rammer and divided with a platinum wire into separate discs, which are placed in glass boxes, moistened with a few drops of milk, and sterilised, and then serve as a nutrient medium on which the most various micro-organisms grow in a characteristic manner, particularly those which form pigments.

The milk rice was modified by Eisenberg in the following way:—100 grm. rice powder, 70 grm. bouillon, and 210 grm. milk are rubbed up and introduced into glass boxes and the mass heated on the water-bath until it solidifies. The boxes having been closed and heated on three days in succession in the steam steriliser, a medium of the colour of *café au lait* is obtained, on the smooth surface of which the micro-organisms can be inoculated.

Bread pap is prepared in the following manner:—Bread is dried until it is fairly free from water, and is then crumbled to powder and spread over the bottom of a flask so as to cover it. By adding water a pap is formed which after boiling yields without being neutralised an excellent medium for the cultivation of *moulds*, the *Aspergillus niger* developing particularly well upon it. After neutralisation with sodium carbonate the pap forms a nutrient medium for different bacteria when it has been several times sterilised in the steam apparatus.

Wafers, especially the thicker ones, are, according to Schill, to be chosen for *chromogenic* bacteria. They are moistened with bouillon, laid in glass boxes, and sterilised.

Modes of Cultivation

Slide cultures.—Microscopic slides covered with a layer of gelatine have repeatedly been used to catch the micro-organisms deposited from the air in different places, or have had a linear inoculation made on the surface. These slide cultures are less in use at the present time, since, in order to isolate the germs, a large surface must be given to the nutrient medium, and consequently glass plates are employed. Moreover, Koch has devised an exceedingly ingenious method known under the name of the *plate process*.

Koch's plate process.—Gelatine is stored in sterilised test-tubes, plugged with cotton, to the amount of about 10 c. cm. in each, and is completely sterilised by the fractional method. A pure culture, or a mixture of many micro-organisms in a mass of any desired size having been obtained, the gelatine in three test-tubes is liquefied in the water-bath at a temperature of 35° C., and a small quantity of the mass to be examined is taken on a platinum needle (previously sterilised at a red heat) and introduced into the first test-tube. The platinum needle may be bent round into a circular loop at the end or have its point somewhat flattened out. If the seed mass is rather too coherent, attempts must be made to separate the micro-organisms by rubbing them with the point of the needle against the side of the test-tube below the surface of the gelatine. The platinum needle having been again heated, three samples are transferred with it from the first tube to the second, and the same procedure is repeated with the second and third, so that we have three inoculations, of which the third is the most diluted. When inoculating care must be taken in opening the tube to seize the plug of cotton-wool between the fingers, best between the third and fourth, on the back of the hand, and thus twist it out of the tube, which must again be carefully closed after inoculation without allowing the cotton plug to come in contact with the surface of the hand or with any instrument.

THE PLATE PROCESS

During these operations some plates should be undergoing sterilisation in the hot-air steriliser at a temperature of 130° C. A box of sheet iron may be used with advantage for this purpose, and is, moreover, capable of containing a larger number of plates (fig. 24). After cooling, three plates, which must only be seized by their corners, are next taken out of the box and laid one after the other on Koch's plate-making apparatus, on which they are cooled under a bell-glass. In the absence of a hot-air steriliser the plates may be sterilised in the interior of an oven, or in the gas or spirit flame by holding them by the corners in the fingers and heating both sides over the flame.

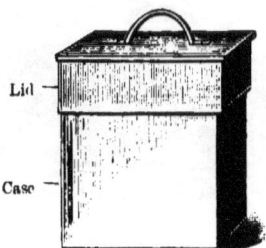

FIG. 24.—CASE OF SHEET-IRON FOR HOLDING THE PLATES.

The *plate apparatus* (see p. 27) consists of a triangle with feet formed by levelling screws, on which rests a glass vessel covered with a thick plate of glass and filled before use with iced water. It is rendered horizontal with the aid of a spirit-level and covered with a bell-glass. The sterilised plates having cooled under this bell, the first of the inoculated test-tubes is then unplugged, its upper edge is heated in order to sterilise that part over which the gelatine has to flow, and its contents are poured from a small height upon the plate in such a manner that the gelatine spreads out over it in a fairly thin layer. In a short time the gelatine sets under the bell-glass, and the plate is then brought into a moist chamber and laid upon either little glass benches or pieces of glass which have been sterilised. The same process is gone through with the second and third inoculations, and the three plates can be laid on benches one above the other in a single moist chamber, or a separate one appropriated to each of them.

The *moist chamber* consists of a large glass box, which is disinfected with corrosive sublimate solution and has a circular piece of blotting-paper moistened with a 1 per 1,000 solution of the same substance laid upon the bottom (fig. 25).

The plates prepared as above are left at the temperature of an ordinary room until the individual cultures show themselves on the surface. These appear in the form of islets

FIG. 25.—MOIST CHAMBER.

either lying close together or isolated. Sometimes several run into one another, and at times a dotted mass appears on the plate, not unusually in the form of little clouds, all of which vary in figure according to the kind of micro-organism. Under a moderately high power of the microscope the colonies are seen to be sharply defined, and sometimes granular, sometimes fibrous, according to the manner in which the micro-organisms are arranged in relation to one another. If the microbes under observation are pigmented, the individual colonies will appear of various tints, or the colour may be diffused through the gelatine, and phosphorescence or fluorescence may be seen in single spots. By comparison of all these peculiarities it is possible to isolate micro-organisms and to identify them, and a further point is that some microbes liquefy the solid gelatine, while others leave its consistence unchanged (fig. 26).

The colonies of micro-organisms are best isolated in the most diluted of the three cultures.

If it is wished to obtain further cultivations from such a plate, the following is the course adopted :—A sample is taken from a colony with the point of a platinum needle fused into a glass rod (the needle being first sterilised at

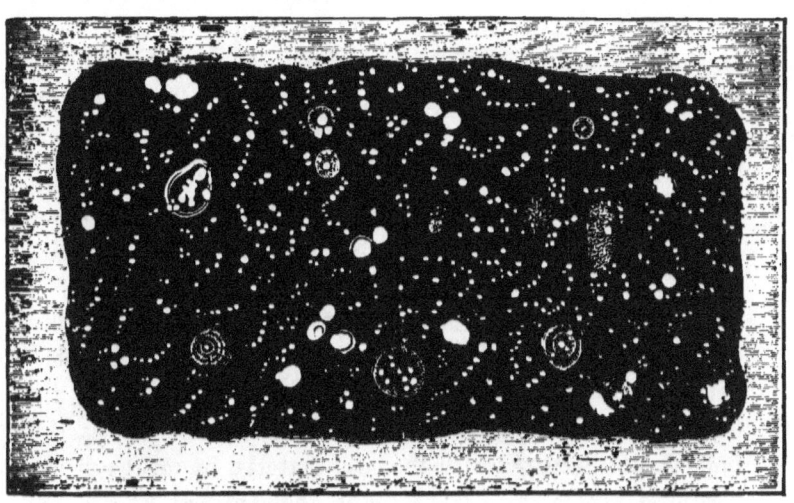

FIG. 26.—GELATINE PLATE, SHOWING COLONIES OF VARIOUS FORMS.

red heat), or the whole colony is lifted on the point. In either case a thrust is made into the sterilised gelatine in a test-tube, or a streak is drawn over the oblique surface of the solid agar-mass, or sterilised potatoes are infected. Such a transference to different nutrient media enables us to note all peculiarities in growth, and hence to gain an inkling of the class in which the micro-organism under consideration is to be included. This procedure is carried out under a low power, and is designated 'fishing.' It requires a certain amount of skill, and therefore Unna, Fodor, and others have quite recently described contrivances for facilitating the operation.

Roll cultures.—A modification of the plate process which is known as *roll culture* has been invented by Von Esmarch, in which the gelatine, after being liquefied and inoculated, is kept rolling round the sides of a wide test-tube until it sets.

Roll cultures are prepared by inoculating the tube in the usual way, closing it with a cotton-wool plug which has first been singed in the flame, and drawing over the cotton an india-rubber cap sterilised in solution of corrosive sublimate. The tube is then seized by its upper end with three fingers of the right hand and by its lower with three fingers of the left, laid horizontally in a vessel of iced water, and kept turning on its long axis until the gelatine has set in a layer of even thickness. The roll cultures when finished must at once be conveyed into a cool place.

Modifications of the plate process.—A quicker and more convenient procedure is the introduction of portions of the gelatine into Petri's capsules. In using Soyka's plates a small quantity of liquefied gelatine is deposited in each hollow, and prepared with the seed-material by means of a platinum needle. By transferring the material from one hollow to another, it is possible to have all degrees of attenuation on the same plate.

In order to economise gelatine Günther places on a sterilised surface of glass a few drops of sterilised water or bouillon, lying isolated from one another. A sample of the material to be examined is mixed with the first drop by means of the platinum wire, and by the same means the inoculating matter is transferred to the remaining drops one after the other, the needle being continually sterilised at a red heat after each inoculation. From the last drop a loopful is conveyed into a test-tube of liquefied gelatine, which is then poured out into a Petri's capsule.

Agar can be used instead of gelatine for the plate process, but this medium requires greater care in preparing the cultivations than is the case with gelatine. Agar becomes fluid at 90° C., and passes into the solid state at 40° C., hence the liquefied mass must be cooled down to 40° before it is inoculated, since at a higher temperature the microorganisms might be destroyed. The mass when inoculated is poured out upon sterilised plates with the precautions mentioned above, and as the water of condensation which separates out renders the film of agar liable to slip from the plate, it is prevented from doing so by dropping some sealing-wax on the edge of the medium. But a more convenient plan is to use Petri's capsules or Soyka's plates to contain the mass. Agar plates have the special advantage that they can be kept for a considerable time at incubation temperature, and that they do not undergo liquefaction.

The roll process can also be applied to cultures on agar.

To facilitate the isolation of micro-organisms, Dahmen has devised an apparatus consisting of a double capsule, of which the upper part extends beyond the lower. The latter is placed on a glass plate and surrounded with an india-rubber ring, on which the upper capsule rests securely. The lower capsule is prepared with the inoculated nutrient agar, placed in the centre of the rubber ring, and covered over with the larger capsule; the whole is then surrounded with an india-rubber band and deposited in the incubator.

The individual colonies form on the agar plate in the same way as on the gelatine (except that they do not liquefy it), appearing coloured or glossy, and showing characteristic outlines.

Plate cultures on serum and plover's egg albumen.—Blood

serum is only used in the solid state, and is principally adapted for surface or *streak* cultures (*Strichculturen*). In order, however, to render it available for plate cultivation, Hüppe mixes it with an equal quantity of a warm solution of agar; and it may also be suspended in gelatine. Unna increased the coagulability of blood serum by rendering it strongly alkaline, in order to add it to gelatine or agar. This medium has special excellences for a number of micro-organisms.

The albumen taken from plovers' eggs which have been previously sterilised with 1 per 1,000 corrosive sublimate may be inoculated and then dried over sulphuric acid under the receiver of an air-pump. Plates so inoculated can be kept for a considerable time if preserved in a sterile place, and when laid in a moist chamber there appear on the thin dry transparent film of albumen variously shaped areas of different sizes, which may be transferred to other culture media. Plover's egg albumen may also, like blood serum, be mixed with gelatine or agar, and so used for plate cultures.

The micro-organisms can be transferred to other media from agar and serum plates in the same way as from gelatine, and the appearances presented by them in their growth observed.

Cultivation of anaerobic micro-organisms.—For the cultivation of *anaerobic microbes*, that is, those which grow with a scanty supply of oxygen, or when it is totally excluded, an entire series of methods have been devised, of which the following are a few.

The most direct plan is to introduce at once into the nutrient medium such substances as will extract the oxygen present, a result which is attained by adding to nutrient gelatine 2 per cent. of grape sugar or 0·1 per cent. of resorcine, or to liquefied nutrient agar $\frac{1}{2}$ per cent. of formic acid or of sodium sulphindigotate.

CULTIVATION OF ANAEROBES

In order to prepare *plate cultures of anaerobic microorganisms* the oxygen can be excluded after Koch's method by laying upon the gelatine or agar before it has fully set a thin sterilised plate of mica or selenite, which adheres closely to the surface of the nutrient mass. The exclusion of oxygen is rendered complete if melted paraffin be run round the border of the mica plate.

The removal of oxygen is effected by Buchner in a very simple way by means of an alkaline solution of pyrogallol, prepared by dissolving a gram of pyrogallol in 10 c.cm. water, and adding 1 c.cm. concentrated solution of caustic potash.

Another mode of cultivating anaerobic microbes on plates consists in bringing the plate prepared with the micro-organism under the receiver of an air-pump and expelling the oxygen by pumping.

Blücher and Botkin secure the removal of oxygen by displacing the air in the receiver with another gas, viz. hydrogen, by means of an india-rubber tube, the lower opening of the receiver being closed with paraffin or with glycerine and water.

Complete *removal of the oxygen from a test-tube* by pumping has been effected by Gruber in the following way:—A test-tube of more than the usual length is drawn out at about 15 cm. from the bottom to a narrow neck. It is filled with 10 c.cm. of nutrient material by the aid of a funnel, closed with cotton-wool and sterilised. After inoculation, the wool plug is pressed down as low as the narrowed part and a tight-fitting rubber cork introduced into the mouth, through a hole in which passes a right-angled tube of glass connected with an air-pump. The air is then pumped out (the culture medium in the rarefied space being meanwhile kept in a water-bath at 30°–40° C.), after which the tube is sealed at the constricted part over the flame of a Bunsen burner, and the gelatine rolled out by Esmarch's method.

In order to displace the air in a test-tube by means of hydrogen, Fuchs recommends that the inoculated tube should be inverted and hydrogen conducted into it from below through a glass pipe, after which the test-tube is closed with a rubber cork.

For test-tube cultivation, however, Liborius' method of preparing *high cultures* is specially adapted. A tube is filled high up with gelatine or agar, which is then freed from air and oxygen by thorough boiling and is cooled to 40°; the matter to be inoculated is distributed in it as evenly as possible with a platinum needle, and it is made to set rapidly in iced water. By this means the deeper layers of the nutrient mass are protected from the air by those higher up, while the superficial ones are exposed to the action of oxygen. When several varieties of bacteria develope, a means is hereby afforded of distinguishing the aerobes which grow on the surface, from the anaerobes growing in the deeper parts.

High cultures are also used for obtaining *thrust* cultivations (*Stichculturen*) of anaerobic micro-organisms, the platinum needle charged with infecting matter being thrust as deeply as possible into the stiff nutrient mass. Although at first development only occurs in the deeper parts, the growth gradually mounts upwards as the gaseous products of its metabolism displace the air from the higher layers of the medium.

Nikiforoff cultivates the anaerobes in the '*hanging drop*.' A cover-glass is prepared with an inoculated drop of bouillon and sealed to a hollowed slide with a layer of vaseline. Between the edge of the cover-glass and that of the well in the slide, the contents of a platinum loop of strong solution of pyrogallol are allowed to flow in on one side, and on the opposite, after pushing aside the cover-glass, a similar quantity of caustic potash, so that the two fluids mix when the object has been brought into the right position.

Hens' eggs seem to afford a suitable medium for anaerobic cultures. This kind of cultivation, which is recommended by Hüppe and Heim, has been described more in detail above (see p. 50).

In cultivating *obligate anaerobes* the materials used in the investigation must, according to Kitisato, be previously heated, in order to remove by this means the facultative anaerobes.

Permanent cultures.—To preserve cultivations of bacteria so that they can be examined at any time, evaporation of the moisture contained in the nutrient medium must be prevented, as well as all possibility of contamination, and consequently the vessels must be hermetically closed. Král punches cylinders out of boiled potatoes, cuts them into discs, and places them in round glass boxes, the covers of which are tightly ground on and the channels in them filled with glycerine. After inoculation they are closed germ-tight with paraffin and spirit varnish. Cultures may also be preserved in test-tubes hermetically sealed by melting the glass.

Prausnitz pours an aqueous solution of gelatine containing 1 per cent. of carbolic or 5 per cent. of acetic acid over thrust cultures placed in iced water, and then closes them with corks and seals them.

By Duclaux's method the cultures of bacteria are enclosed in the small tubes used to contain lymph.

Jacobi first exposes the gelatine which contains the colonies, and which has been spread out in the thinnest possible layer, to the action of 1 per cent. bichromate of potash for from one to three days in the presence of light, hardens it in alcohol, and cuts it into pieces which can then be strained like sections of tissue, and thus rendered permanent.

To preserve small portions from agar plates, Günther lays them in a drop of glycerine placed on a slide, deposits thereon a second drop, and finally the cover-glass. The superfluous glycerine having been absorbed out the preparation is closed with cement.

CHAPTER IV

EXAMINATION OF MICRO-ORGANISMS UNDER THE MICROSCOPE,
AND BY EXPERIMENTS ON LIVING ANIMALS

Examination in the fresh state.—In making microscopic examinations we begin with the simplest mode of procedure, which consists in taking minute samples from the individual colonies on the plate with the help of the platinum needle, floating them in water, and subjecting them to observation. It must be seen to that too much fluid is not taken, but only enough to fill the interspace between cover-glass and slide. The former ought not to float about loosely, nor should the fluid extend beyond its edges. When examining with high powers it must be noted which *form* the particular micro-organism takes—that is, whether rods or cocci are to be dealt with, whether they are connected with one another in chains, whether, if so, the chains run straight or spirally, and, in the case of cocci, whether they lie scattered or in rows. *Size* is measured in micromillimeters (*micra*, commonly written μ, = the thousandth part of a millimeter), or by comparison with other similar forms, especially red corpuscles.

Examination in the hanging drop.—A very useful method of observing freshly-obtained micro-organisms is the examination in the '*hanging drop*.' For this purpose the end of a platinum wire is bent with the aid of a pliers into a little loop. When this is dipped into a liquid containing bacteria, enough of the liquid remains adhering to it to form a small drop, which is transferred to a cover-glass. A 'hollow' slide is then taken, that is, one with an excavation or well ground in the centre; this well is surrounded with vaseline by means of a fine hair pencil, and the cover-glass with the drop turned downwards is laid on the slide in such a way as to adhere firmly to the vaseline. If the substance, the microbic contents of which it is desired to examine, is not a liquid containing bacteria, but animal tissue or solid culture medium, a drop of sterile water or sterile salt solution is conveyed on to the cover-glass with the loop of the platinum needle, and a minute sample of the mass to be examined is transferred into the drop.

In observing the hanging drop the edge must first be sought for with a low power, and then focussed with a higher; since, as it appears bounded by a sharply-drawn line, the micro-organisms in the drop can in this way be more easily focussed, which very much facilitates the examination for beginners; and, morever, the elements are met with in a thinner layer at the border than in the centre. As the elements under observation are not stained, the narrowest possible aperture of the diaphragm must be used in the examination.

With the hanging drop attention must in like manner be paid to the peculiarities which can be observed in bacteria examined in the fresh state, as detailed above, and their motility is more distinctly brought out in this mode of investigation. The closure of the space prevents the fluid from evaporating, but if the examination is too prolonged the micro-organisms sink into the concavity of the drop, and so sometimes elude observation.

When it has been ascertained by means of fresh preparations that micro-organisms are present, and their form, mode of propagation, and power of movement have been observed, the next step is that of staining. So many processes have been established in the course of the researches which have been made up to the present that we must describe them here more in detail, especially as they afford marks by which bacteria can be distinguished, and consequently staining is of the highest importance in determining the individual varieties.

Staining of micro-organisms.—Staining constitutes an indispensable aid to the study of the finer structure of the micro-organisms, and of their relation to the cells of the body. In carrying it out a series of colouring matters are used which have been already detailed (p. 30), and solutions are prepared from these in different ways, which are employed both for isolated micro-organisms and also for those in the tissues. The basic aniline colours are for the most part kept in stock in alcoholic solutions which are mixed with water before use, so that we really employ a dilute alcoholic solution in staining. The dilution, however, must not be carried too far.

Günther has pointed out that absolute alcohol is not suitable for use in staining with basic aniline colours, just as it is incapable of extracting the dye from cells when once they have been stained.

Simple staining of cover-glass preparations.—In this, the simplest kind of staining, the mode of procedure is as follows: A sample of the matter to be examined is conveyed on to a cover-glass with the point of the sterilised platinum needle and is diluted, if needful, with water, after which the organisms suspended in the water are spread out over the surface of the glass with the flattened end of the needle (*smear preparation*); or a better way of managing this is to press another cover-glass upon the prepared one, and then slide it off, so that the mass under examination appears equally distributed on both cover-glasses. The mass which contains the micro-organisms is frequently found to have so much moisture as to render the addition of water superfluous. If it is desired to examine the juice of organs, a piece of the organ is seized in a forceps and the cover-glass smeared with it, dried in the air, and passed three times through a flame to *fix* the micro-organisms to its surface, after which the staining is done by depositing a few drops of dye on the infected surface of the cover-glass, or by pouring some into a watch-glass and floating the cover-glass upon the solution with the prepared side downwards. After from one to five minutes it is freed from superfluous stain by washing in water, is dried in the air —a process facilitated by soaking up the drops with blotting-paper—and is mounted in fairly fluid Canada balsam; or, if it is not wished to preserve the preparation, it may be examined in water, or in a very dilute solution of potassium acetate. Such objects are examined with ordinary or homogeneous immersion objectives by the aid of Abbé's illuminating apparatus without a diaphragm. Coloured preparations admit of being seen with distinctness, and their outlines can be accurately determined, such figures being spoken of as 'coloured images,' to distinguish them from the unstained 'structural images,' which should only be examined with a diaphragm of narrow aperture. When Abbé's apparatus is used without a diaphragm, all the rays which enter the lower lens, and which form a very obtuse-angled pencil, are enabled to reach the object.

Bacteria are difficult to observe in fluids and tissues, being only visible through the shadows caused by the differences in refractive power of the several structures. Hence but little light must be allowed to reach the preparation, and consequently as small a diaphragm as possible used, and the result is an impairment of distinctness. If, however, the bacteria be stained, it becomes possible to remove the diaphragms, and to examine with the full power of the Abbé's illuminator.

The coloured image is best if the structural image be effaced by rendering the shadows of the unstained parts invisible in the broad cone of light.

It must further be remarked regarding the structural image, that the diaphragm should have the narrowest possible aperture with a low power, but should increase in size as higher powers are employed.

Preparation of staining solutions.—For the simplest kind of staining of bacteria solutions of fuchsine, methyl blue, gentian violet, Bismarck brown, vesuvine (in equal parts of water and glycerine), and methyl violet are used. Gentian violet and fuchsine stain quicker and more intensely than the others. In order to increase the staining power with the various micro-organisms, certain *mordants* are employed, as in other histological methods of staining. *Aniline oil* and *phenol* are the mordants most used in bacteriological research. The former, which is not a true oil, is obtained from coal tar, and it is used for preparing an *aniline water* in which the dyes, especially gentian violet or fuchsine, are dissolved. The aniline water should be prepared freshly each time, or in any case should not be allowed to stand long, as it rapidly decomposes. To make it a test-tube is filled with water which is shaken up with 1 to 2 c.cm. of aniline oil until an emulsion is formed, which is filtered. The clear filtrate is aniline water ready for use, and enough of the alcoholic solution of the dye is then added to render the liquid of a dark colour. The concentration of aniline water amounts to 5 to 100.

Trenkmann prepares his *aniline water solution* of gentian violet in the following way:—A drop of a concentrated alcoholic solution of gentian violet is let fall into a test-glass and 10 c.cm. of water are added. Half of this is then poured away, and the glass filled with aniline water; a solution is thus obtained which remains clear and stains the bacteria themselves deeply, but the ground very slightly. The cover-glasses should remain about half an hour in the staining fluid.

Instead of the aniline water a 5 per cent. aqueous solution of *carbolic acid* (*phenol*) may be used, to which an alcoholic solution of fuchsine is added, until the mixture becomes of a dark colour. This mixture is known as *Ziehl's solution*, and its composition is as follows:

Crystallised carbolic acid	5·0
Water	100·0
Alcohol	10·0
Fuchsine	1·0

According to Kühne's formula, methyl blue instead of fuchsine is mixed with 5 per cent. carbolic acid, water, and alcohol, and a solution of strong staining power so obtained.

Instead of carbolic solution a 1 per cent. solution of *ammonium carbonate* can be used as a mordant.

Koch employs a solution of *caustic potash* as an ingredient, adding to 1 c.cm. of a concentrated alcoholic solution of methyl blue 200 c.cm. of water and 0·2 c.cm. of a 10 per cent. potash solution, and Löffler also adds 100 c.cm. of 0·01 per cent. solution of caustic potash to 30 c.cm. concentrated alcoholic solution of methyl blue.

In certain staining processes, particularly that for tubercle bacilli, a *sulphuric acid* solution of methyl blue is employed, prepared by mixing 100 parts of a 25 per cent. solution of sulphuric acid with 2 parts methyl blue; or a *nitric acid* solution consisting of a saturated solution of methyl blue in 20 parts nitric acid, 30 parts alcohol, and 50 parts distilled water.

The different methods of staining are in many cases assisted by heat, the colouring solutions being kept warm on the cover-glass or in watch glasses while the process is going on. When, however, sections of tissue containing bacteria are to be warmed, precautions must be taken to avoid spoiling the tissue, especially as staining takes longer in the case of sections.

Staining of flagella.—For the purpose of rendering visible the flagella of motile micro-organisms, Löffler uses a mixture of 10 c.cm. of a 20 per cent. solution of tannin and a few drops of saturated ferrous sulphate solution, with fuchsine or 4 or 5 c.cm. of extract of logwood. Staining is effected with fuchsine in aniline water, to which a 1 per 1,000 solution of caustic potash has been added until it becomes turbid owing to a floating precipitate. For bacteria which form alkalies, the mordant must be rendered correspondingly acid; for those which form acids, alkaline.

According to Trenkmann the cilia are brought into view if the preparations are treated before staining with tannin and hydrochloric acid or catechu tannic acid to which carbolic acid has been added, or extract of logwood treated with acid; and they become still more distinct if the preparations, after being treated with the mordant and stained, are examined in a drop of iodine water. Two or three drops of boiled water are allowed to fall upon a slide, and a small drop of the culture to be examined is added and mixed well in. A minute droplet is conveyed from this to a cover-glass, spread out, dried in air, and laid without previous heating in a solution containing 2 per cent. of tannin and $\frac{1}{2}$ per cent. of hydrochloric acid. The cover-glass remains for from 6 to 12 hours in this solution, and is then washed in water, laid for an hour in iodine water, washed again, and deposited for half an hour in a weak solution of gentian violet in aniline water.

Staining of spores.—It is possible to effect a staining of the spores, in those micro-organisms which form them, by warming the staining fluids. Carbolic acid fuchsine (*Ziehl's solution*) is used for staining, and the infected cover-glass is left for an hour in the boiling dye, when the spores in the bacilli will remain of a red colour after washing with water and decolorising with alcohol; or if double staining with methyl blue is carried out the bacilli appear blue, and the spores dark red. The *hay bacillus*, and especially the *Bacillus megatherium*, should be selected for the study of these methods of staining. In many of the microbes hitherto known the discovery of spores has not as yet been made.

Spores are also brought out distinctly as little granules by staining with dilute alkaline methyl blue, in which case, after double staining with aqueous solution of Bismarck brown, they appear blue on a brown ground.

According to Moeller, spores are most conveniently stained by the following method :—The cover-glass preparation is brought for two minutes into absolute alcohol and for two more into chloroform, washed in water, plunged for from a half to two minutes into a 5 per cent. chromic acid solution and rinsed again with water, after which some aqueous carbolic fuchsine solution is dropped on, and it is warmed for one minute in the flame, being brought once to the boil. The carbolic fuchsine is poured off, the cover-glass dipped into 5 per cent. sulphuric acid until decolorised, and once more thoroughly washed with water. Finally, an aqueous solution of methyl blue or malachite green is allowed to act on it for half a minute, and then washed off. The spores are dark red in the interior of blue or green bacteria.

Decolorising Agents

In staining with the various aniline dyes a phenomenon of practical importance has been observed, viz. that stained micro-organisms part with their colouring matter to certain reagents. These are known as *decolorising agents*, and include amongst them *water, alcohol, acetic, hydrochloric, sulphuric*, and *nitric acids, iodine*, &c.

Upon decolorising with one or other of the above-named fluids depend various methods which have acquired an extraordinary significance for the diagnosis of micro-organisms, and for the practical work of staining them in sections.

Koch and Ehrlich method of staining. — The Koch-Ehrlich method of staining tubercle-bacilli stands in the first rank of these, and brings into action both the mordant and bleaching processes. Aniline water is prepared as described above, and alcoholic solution of fuchsine, gentian violet, or methyl violet is added to it until a fairly saturated solution in dilute alcohol is obtained. Small masses of sputum, or of the wall of a pulmonary cavity, are then conveyed on to a cover-glass and spread out by rubbing with a second, so that both glasses become coated with a fine film of the mass under examination. The cover-glasses, having been dried in air, are passed three times through the flame with the prepared side up by means of a forceps, and then deposited in the staining fluid and either left for twenty-four hours at the temperature of an ordinary room, or heated for fifteen minutes until bubbles rise. The cover-glass is next lifted from the dye with a forceps and plunged for a few seconds into a solution of about 33 per cent. of nitric acid until the preparation, previously of a red colour, becomes yellowish green, and is then washed in 70 per cent. alcohol. If fuchsine has been used for the staining, the after-staining may be done with methyl blue, malachite green, or picric acid; but if the first staining has been done with gentian or methyl violet, Bismarck brown must be used for the second. The secondary staining lasts from one to five minutes, until the particular colour used is plainly visible on the cover-glass, after which this is washed in water, dried, and mounted in Canada balsam. After washing in water the cover-glasses may also be brought into alcohol and then treated with oil of cloves and Canada balsam.

In this staining process the nitric acid acts as a bleaching agent on the different micro-organisms contained in the mass under examination. The tubercle-bacilli alone refuse to yield up their stain to the acid unless it has acted for a long time.

Ziehl and Neelsen's method of staining.—The Koch-Ehrlich method was modified by Ziehl and Neelsen by using carbolic fuchsine instead of aniline water fuchsine. Here, also, either the stain is applied on the cover-glass, or this is laid prepared side downwards in the warm dye. It is afterwards washed with water and decolorised in 33 per cent. nitric or 5 per cent. sulphuric acid. In all other particulars the process resembles the Koch-Ehrlich method, and secondary staining is effected by means of malachite green, picric acid, or methyl blue.

Ehrlich's method of staining.—For demonstrating tubercle-bacilli in pus, Ehrlich recommends that it be spread out very thinly, and the preparation placed for one to two hours in cold aniline fuchsine and decolorised with sulphanil-nitric acid (1 part nitric acid to 3 to 6 parts saturated solution of sulphanilic acid). The double staining is done with methyl blue.

Günther's method of staining.—In this process the staining is effected with warm aniline water fuchsine, from which the cover-glass is conveyed with the prepared side uppermost into alcohol containing 3 to 100 of hydrochloric acid, in which it is moved about for a minute and then rinsed in water. By means of a pipette a few drops of dilute alcoholic solution of methyl blue are now allowed to fall upon the cover-glass, which is then washed in water, dried, again passed three times through the flame, and mounted in Canada balsam and xylol.

Weichselbaum's method.—This is a modification of the Zichl-Neelsen method, in which the red-stained cover-glass preparations are transferred directly to an alcoholic methyl blue solution, in which they remain until they show an even blue colour. They are then rinsed in water, dried, and mounted in Canada balsam. The alcohol is here the only bleaching agent.

Fraenkel's method.—The cover-glasses are stained with aniline water fuchsine and transferred to a fluid consisting of a saturated solution of methyl blue in 50 parts water, 30 parts alcohol, and 20 parts nitric acid. When the preparation appears blue it is washed in alcohol and acetic acid or in pure water, and is examined in water.

Gabbet's method.—After staining in carbolic acid fuchsine the preparation is brought into a sulphuric acid methyl blue (2 grms. methyl blue to 100 grms. of a 25 per cent. solution of sulphuric acid), and double stained with malachite green.

Method of Pfuhl and Petri.—The preparations are stained in a mixture of 10 c.cm. alcoholic solution of fuchsine to 100 c.cm. of water, and subsequently decolorised in glacial acetic acid. They are then washed in water and double stained with malachite green, again washed in water, dried, and mounted in Canada balsam. In this case the glacial acetic acid is the decolorising agent.

Method of Pittion.—The cover-glass preparation is dipped for a minute into a mixture of 1 part alcoholic fuchsine solution with 10 parts of a 3 per cent. solution of ammonia, and transferred after rinsing in water to a concentrated solution of aniline green in 50 grms. alcohol, 30 grms. water, and 20 grms. nitric acid, in which it remains ¾ minute.

Arens' chloroform method.—In order to avoid heating, as well as the preparation of a complicated staining fluid, an alcoholic solution of fuchsine mixed with chloroform is used as the dye, and alcohol with hydrochloric acid for decolorising. The fuchsine solution is prepared by pouring 3 drops of absolute alcohol on a crystal of fuchsine the size of a millet-seed in a watch-glass and adding 2 to 3 c.cm. chloroform. The solution becomes turbid and then begins to clear, flocculent particles of fuchsine being separated. When the clearing is complete the cover-glass preparation is laid in it for from 4 to 6 minutes, until the chloroform is evaporated, and is then decolorised in concentrated alcohol, to which hydrochloric acid has been added in the proportion of 3 drops to a watch-glass full. It is next rinsed in water, and finally double stained with dilute methyl blue.

Gram's decolorising method.—This appears to be the most extensively used of all the bleaching methods, and depends upon the employment of iodine in aqueous solution combined with potassium iodide (1 part iodine, 2 parts potassium iodide, and 250 parts water) after the preparation has been stained in aniline water solution of gentian violet. The iodine forms with the colouring matter a precipitate which adheres to the micro-organisms, but can be easily washed out of the tissues, and if this is properly done the bacilli or cocci appear isolated by the stain. The following is the mode of procedure in carrying out Gram's method :—The prepared cover-glass or section containing bacteria is warmed in aniline water gentian violet; too strong heating, however, has an injurious effect. The best mode of warming the solution is to place it for 15 minutes on the water-bath; or to hold it over the flame for 1 minute, let the watch-glass cool for 3 minutes, then heat it again for a minute more and let it cool again for two or three, and so on, until the process has been repeated four or five times.

The preparation is next laid for 1 to 2 minutes in the iodine and potassium iodide solution and transferred from that into absolute alcohol, in which it remains until the colour is discharged. The bacteria come out stained with gentian violet, and the tissue may be double-stained red with picrocarmine, Magdala red, or other dyes.

Gram's method can be used as an aid to the diagnosis of the vast majority of micro-organisms. For example, the *Pneumococcus Friedländer* shows no staining after going through the process, and similarly the bacilli of *Cholera Asiatica, typhoid fever* and *glanders, gonococci*, the *spirilla of recurrent fever*, &c., cannot retain the colouring matter, but give it up, as do also the nuclei of cells, when iodine solution is applied.

It is strongly to be recommended that the preparation should not be brought directly from the staining fluid into the iodine and potassium iodide, but be first rinsed free of superfluous stain in plain aniline water before being transferred to the iodine solution (Botkin).

In staining sections of tissue it is advisable to carry out the ground staining before that of the bacteria, which is done by immersing the sections in picrocarmine for one or two minutes, washing in water, transferring to alcohol, and then subjecting to Gram's process.

Every pigment is not, however, suitable for this method, since Unna has shown that it gives no results if fuchsine, methyl blue, or Bismarck brown are used. The process can only be carried out with the pararosanilines (*methyl violet, gentian violet*, and *Victoria blue*).

Günther's modification of Gram's process.—Not only pure alcohol, but also alcohol to which 3 per cent. of hydrochloric acid has been added, is used for decolorising. The cover-glass or section of tissue is left for about two minutes in aniline water gentian violet, but in the case of *tubercle* *bacilli* the dye is allowed to act for twelve hours, and in that of *lepra bacilli* for half a day. Superfluous stain is removed with blotting-paper, and the preparation is brought for two minutes into solution of iodine and potassium iodide, then for half a minute into pure alcohol, for exactly ten seconds into 3 per cent. hydrochloric acid in alcohol, and immediately afresh into plain alcohol for several minutes, changing the spirit as long as any colour is extracted from the preparation. The cover-glass is now dried and mounted in balsam, and sections of tissue are laid in xylol (which renders them transparent in half a minute), and then mounted on slides with Canada balsam dissolved in the same liquid.

Weigert's modification of Gram's process.—The sections stained with gentian or methyl violet are not transferred to alcohol from the iodine solution, but laid upon slides and covered with aniline oil, which dehydrates and differentiates them. The aniline oil is then removed with blotting-paper, xylol is poured upon the preparation, and it is put up in Canada balsam in xylol.

Impression preparations.—These are made for the purpose of rapidly gaining an idea, when examining plates, regarding the arrangement of the colonies and the microscopic peculiarities of the organism under investigation. A cover-glass is laid on the plate, pressed gently down, lifted carefully with a forceps, and laid aside to dry. It can then be stained like an ordinary cover-glass preparation.

Examination of micro-organisms in sections of tissues.— The examination of micro-organisms in the tissues, whether in the interior of the individual cells, or in the structures which are formed by them, is of pre-eminent importance in research directed to medical ends. Not only have the nature of the micro-organisms and their mode of entrance into the body to be discovered, but attempts must be made to ascertain their bearing towards the elements of the tissues, and their exact situation in and between them. In particular, their physiological action, in spite of very advanced methods of investigation, has not up to the present been fully explained. When the microbes are present in masses of considerable size, but only then, their position in the tissues can be recognised in the simplest way in pieces of the fresh organ by examining some of its tissue on a slide in a drop of sterile water or salt solution. In the case of fluids the addition of water may be omitted. The minute elements are then examined with a high power (an oil-immersion with Abbé's condenser and a diaphragm). We cannot, however, ascertain by this method how the microbes are related to the tissues, nor their exact situation in the tissue and its elements. A small piece of it should therefore be torn up with needles and treated with a drop of acid or caustic potash, so as to cause the connective tissue which forms the bulk of organs to swell up. In this way the bacteria, which exhibit greater power of resisting reagents, are brought out distinctly. Through the introduction of staining processes, however, methods have been discovered which render it possible to demonstrate the micro-organisms in uninjured tissue.

Examination by the freezing method.—In order to be able rapidly to examine pieces of organs, recourse is had to the *freezing microtome*. The substance in the fresh state is laid upon a roughened metal plate and frozen by means of an ether spray apparatus. It is then cut into sections with a cooled knife, and these are laid on slides, allowed to thaw, and subjected to staining processes which will be described later on; after which they are most conveniently examined in dilute glycerine.

Hardening.—Owing to the frequent destruction of the tissue in using the freezing microtome, caused by the crystals of ice which form, the organs are usually not cut until they have undergone a hardening process.

In Histology an entire series of reagents is employed for hardening, the use of which is, however, impracticable in Bacteriology, because they deprive bacteria of their property of taking up aniline colours easily. The most convenient way is to harden the pieces, which should each be a cubic centimeter in size, singly in *absolute alcohol*, which must be changed several times. The alcohol may be obtained as free from water as possible in the following way: Crystals of copper sulphate (blue vitriol) are heated in an iron capsule with frequent stirring until they have completely parted with their water of crystallisation and subsided to a white powder, which, after cooling, is introduced into a bottle and the alcohol is poured over it, when it greedily extracts the water therefrom, becoming again blue. As the piece of tissue contains water, it sinks to the bottom if thrown into absolute alcohol, and the hardening process goes on more slowly in the lower than in the upper half of the vessel, the alcohol above being less rich in water. Hence it is advisable to keep the organ in the upper part of the alcohol, either by means of a layer of cotton wool, or by suspending it with a thread fastened outside.

A half per cent. *chromic acid* solution, with or without the addition of *platinum chloride and acetic acid*, has also been recommended, as in it the bacteria are well preserved. After eight days the pieces are rinsed in water until that which flows away shows no yellow coloration, and the hardening is then completed in alcohol.

Instead of chromic acid, a concentrated aqueous solution of *picric acid* renders good service. The pieces are left in this for two days, washed for twenty-four hours in water, and transferred first to dilute, and from that to absolute, alcohol.

Portions of tissue so fresh as to be still warm are best hardened in *corrosive sublimate*. They are left for from ten to thirty minutes in a 5 per cent. solution prepared at 70° C., and are then transferred directly into moderately dilute alcohol, in which they remain for a day, and hardening is then completed in absolute alcohol.

Imbedding.—The hardened specimens are prepared for section-cutting in many different ways, to enable them to be fixed in the microtome.

Imbedding in gum arabic.—One of the simplest methods consists in fastening them with gum arabic to cork or elder pith, or to little bits of wood, when, after drying, sections are cut from them, great care being taken to prevent the hardened gum from injuring the knife. The process consists in immersing the pieces to be cut in a concentrated solution of gum arabic of a syrupy consistence, after imbedding in which they are deposited in concentrated alcohol. This extracts the water from the gum, so rendering the mass sufficiently firm to be cut.

Imbedding in glycerine jelly.—The most useful method is that of attaching the pieces to little bits of cork or wood by means of a concentrated glycerine jelly prepared with the aid of heat; Fränkel recommends boiling together one part gelatine, two parts water, and four parts glycerine. The portion of organ having been made to adhere by means of this glycerine glue, nothing further is done until the gelatine sets, when the piece is laid in alcohol and becomes after some time so firmly adherent that the cork can be clamped in the microtome and sections made. It is necessary to bring the knife to the preparation obliquely, and to keep everything constantly wet with alcohol while cutting. Before staining, the sections must always be brought into absolute alcohol. To enable glycerine jelly to be kept in stock, a drop of corrosive sublimate must be added, in order to prevent the growth of micro-organisms.

Imbedding in celloidine.—The celloidine method is a very convenient one. It consists in fastening the portions of organs to bits of cork or wood by means of celloidine dissolved in alcohol and ether, and then, after the celloidine has set, immersing them in alcohol, in which they gain a consistence suitable for cutting. The pieces are placed in absolute alcohol and left there for twenty-four hours, after which they are transferred to a mixture of equal parts of alcohol and ether, and finally to a celloidine solution of medium consistence, in which they remain for at least twenty-four hours, in order that the tissue may become thoroughly saturated with that substance. The pieces are now taken out one by one and fixed to corks by means of celloidine, and as soon as it has set in the air, which requires only a few minutes, the pieces fastened to the corks are immersed in very dilute (30 per cent.) alcohol. In this the celloidine becomes cloudy after a short time, until after several days it is changed into an opaque white mass of such firmness that the piece of organ to be cut is securely adherent to its cork support, and if this be now fixed in the clamp of the microtome it is possible to obtain the finest sections. These sections are enveloped, so to speak, in a mantle of celloidine, which is capable of taking the aniline dyes.

This method gives good results with Gram's process. When it is wished to stain in this way several sections following one another in series, the *section stainer* [1] in use at the author's Institute is well adapted for this purpose. Several sections having been laid in serial succession upon a slide of larger size than usual, are covered with a nickel-plated grating and clamped in the section stainer. The whole is then passed through the various fluids and stains

[1] Sold by Siebert in Vienna.

one after another, the delicate grating preventing the sections from slipping off, without in any way injuring them; and when finally it is raised after full completion of the treatment, the sections remain lying in their original order, and the result is a *serial preparation* (fig. 27).

Imbedding in paraffine.—This method serves for the preparation of finer sections, but is only rarely used in bacteriology. It is employed for making single sections as well as for series. The pieces of organ are brought into absolute alcohol for twenty-four hours, then into a mixture of chloroform and alcohol for twenty-four more, and finally for the same length of time into pure chloroform. Xylol, oil of cloves, and oil of turpentine do not yield such good results. If the pieces are saturated with chloroform they should sink

FIG. 27.- SECTION STAINER FOR PREPARATIONS IMBEDDED IN CELLOIDINE.

to the bottom in that liquid. After this they are laid in paraffine dissolved by heat in chloroform, and remain in this solution for two or three hours at a temperature of 30°–40° C. Finally they are imbedded in paraffine. Little boxes of paper having been made ready and floated on cold water, fluid paraffine is poured into them, and after it has solidified the pieces of organs are laid upon it and covered with more melted paraffine, which liquefies again the surface of the layer already solidified, so that the specimen seems enclosed in a block of the substance. After a few hours this is trimmed to a suitable form with a knife, clamped in the microtome, and sections are cut with the knife transverse or slightly oblique, and without using any moistening fluid. The microtome can be arranged to cut sections of any desired thickness.

The sections are transferred one by one to xylol, in order to extract the paraffine from them, and are thence brought into alcohol and then into water. If they do not sink in the water the paraffine has not been completely removed, and in this case they must be returned to the alcohol and from that into xylol, and then transferred afresh to alcohol and water. After removal from the water they are subjected to suitable staining processes, cleared in xylol, and mounted in xylol Canada balsam. Oil of cloves should not be used for clearing the tissue, as it decolorises the micro-organisms.

In the preparation of *serial sections* a softer paraffine is used for imbedding; the imbedding-block is otherwise prepared in the same way as before and cut to a square, and the microtome knife is fixed transversely. The sections, which adhere to each other, forming bands which resemble a tape-worm in outline, are laid one beside the other in corresponding order and fixed to the slide, usually with the white of a hen's egg diluted with water and glycerine. At ordinary temperatures the white of egg takes a long time to dry, but this may be expedited by gentle heating. A drop of creosote or carbolic acid should be added to the fluid to make it keep. Other fixing media are *collodion*, *glycerine agar*, or *glycerine gelatine* in a dilute condition.

The adherent pieces are freed from paraffine with xylol, which is extracted in turn with alcohol; they are then washed in water, subjected to staining processes, rendered transparent with xylol in the same way as single sections, and put up in Canada balsam.

On the Staining of Sections

The staining of sections is carried out after various methods, but a certain order of procedure is common to all.

The sections, whether single or in series, are transferred from the alcohol to water, and remain in it until thoroughly saturated, which serves as proof that they are freed from alcohol and from any other fluid that may by chance adhere to them; after which they are subjected to the action of the selected stain for from two to five minutes to twenty-four hours. The time during which the stain must be allowed to act may, however, be shortened by warming, so far as this can be done without spoiling the tissue. The preparation must now be washed in water as long as any colour comes away from it. The various bleaching agents are next used, and from them the preparation is transferred to water, then to alcohol in order to dehydrate it, and is finally cleared with xylol. It is advisable several times to change the alcohol used for dehydrating. *Xylol* is employed because it behaves in a completely indifferent manner towards basic aniline colours, whether in nuclei or bacteria, which is not the case with other clearing reagents; and moreover it evaporates without deposit, never becomes resinous, and consequently does not soil articles with which it comes in contact so much as does oil of cloves. Besides xylol, *oil of turpentine, aniline oil, phenol, oil of bergamot, oil of cedar, oil of origanum, oil of cinnamon*, &c., are used. When the preparation has been rendered sufficiently transparent by means of the xylol, it is transferred to a slide and dabbed with blotting-paper, a drop of Canada balsam in xylol is placed on it, and a cover-glass applied.

Unna's drying-on process (dry method).—Sections cut with the freezing microtome are stained in a dilute alcoholic solution of fuchsine, washed in water, laid for a short time in alcohol, double-stained in methyl blue, dabbed with blotting-paper, dried on a slide over the flame, and put up in Canada balsam and xylol.

Combination of staining methods.—The dyes are selected in the same manner as when staining the bacteria from a plate-culture or from a mixed mass of them; and the combination of several colours is indicated, because then that of the bacteria stands out distinctly from the ground tint of the tissue.

Kühne's methyl blue method.—Kühne, to whom the most marked advances in the technique of staining are due, recommends as the most reliable method the staining of the sections with methyl blue dissolved in a 5 per cent. carbolic acid or a 1 per cent. ammonium carbonate solution. In order to differentiate the preparations, they are brought after staining into a weak aqueous solution of lithium carbonate or into slightly acidulated water, then dehydrated in absolute alcohol to which some methyl blue has been added, and transferred to aniline oil, similarly mixed with methyl blue. Each section is then cleared by immersion in pure aniline oil, next in a light fluid etherial oil, such as that of thyme or terebene, and finally in xylol, and is mounted in balsam.

For staining the bacilli of *tuberculosis*, *leprosy*, and *mouse septicæmia* a method may be used which differs from the foregoing only in the substitution of fuchsine for the methyl blue.

Koch's method.—The sections after staining are transferred to a saturated solution of potassium bicarbonate which has been diluted with an equal volume of water, and thence to alcohol, cedar oil, and Canada balsam.

Löffler's method.—Löffler stains the sections in an alkaline solution of methyl blue, decolorises in half per cent. acetic acid, and thence brings them into absolute alcohol, cedar oil, and Canada balsam.

Chenzynsky's Method.—The sections are immersed in a methyl blue and eosine solution containing forty parts concentrated alcoholic solution of methyl blue, twenty parts of a ½ per cent. eosine solution in 70 per cent. alcohol, and forty parts water, and after staining are rinsed in water, and the remainder of the treatment carried out in the usual way. Plehn recommends the addition of twelve drops of a 20 per cent. caustic potash solution to the water.

Gram's method.—This method is in a high degree suited for sections. They are stained in aniline water gentian violet, to the action of which they are exposed for from ten to thirty minutes; but the time of staining may be shortened by heating. After staining they are rinsed in water and immersed for two to three minutes in a solution of iodine and potassium iodide, and are then kept moving to and fro in 90 per cent. alcohol until no more colouring matter comes away. The sections, which now appear of a slate-grey colour, are next transferred to alcohol, cedar oil, and Canada balsam. The bacteria are seen in violet on a yellowish ground. Double-staining with picrocarmine or Magdala red causes the violet tint of the micro-organisms to stand out distinctly against the red colour of the tissue.

The method of Gram may also be reversed, and the sections *first* stained for fifteen minutes in picrocarmine or Magdala red, rinsed in 50 per cent. alcohol, and then laid in aniline water gentian violet. After decolorising in iodine solution, the preparation is treated with alcohol, oil, and Canada balsam.

Günther's modification, which is characterised by the exposure of the sections, after decolorising in alcohol, to the action of a 3 per cent. solution of hydrochloric acid, yields brilliant results (compare p. 54).

Kühne's modification of Gram's process.—Gram's method has undergone many further modifications in its use for sections, and Kühne in particular has devised a number of processes, of which the following are the most important.

A solution is prepared of 1 grm. *Victoria blue* in 50 c.cm. of alcohol diluted to half its strength, and this is again diluted to the same exten with a half per cent. aqueous solution of ammonium carbonate. Staining lasts from one to five minutes, and the sections are decolorised in iodine and potassium iodide, and further treated as directed by Gram, except that instead of alcohol a solution of fluoresceine (1 grm. fluoresceine to 50 c.cm. absolute alcohol) is used for extracting the colouring matter.

A further process consists in adding some hydrochloric acid (1 drop to 50 grms. water) to a concentrated aqueous solution of *violet* and using this to stain the sections, which are otherwise treated as in Gram's method.

In using *carbolic methyl blue* the sections are stained for from half an hour to two hours, rinsed in water mixed with hydrochloric acid, passed through a weak aqueous solution of lithium carbonate, and transferred from that to absolute alcohol and to aniline oil, in both of which a little methyl blue has been dissolved. After rinsing in pure aniline oil they are cleared in an etherial oil which is then removed with xylol, and are mounted in Canada balsam.

Pragl recommends a modification of the carbolic methyl blue method, which consists in staining the sections, fixed to a slide or cover-glass, for from half to one minute in carbolic methyl blue, after which they are rinsed in water for a short time, decolorised in 50 per cent. alcohol, dehydrated in absolute alcohol, cleared in xylol, and mounted in resin.

Another method which is easily applied consists in staining the sections for three to five minutes in *carbolic fuchsine*, rinsing in water, and passing through alcohol. They are then laid for a quarter of an hour to two hours in aniline oil containing some methyl green, in order to decolorise and differentiate them, and after clearing with an etherial oil and removal of this with xylol, are put up in Canada balsam.

In the *fluoresceine and oil of cloves* method the sections are immersed for five to ten minutes in a concentrated aqueous solution of oxalic acid, which acts as a mordant, rinsed in water, and dehydrated in alcohol. The staining which follows is done with fuchsine in aniline water, or methyl blue dissolved in $\frac{1}{2}$ to 1 per cent. aqueous solution of ammonium carbonate. To dehydrate, the sections are left for five to ten minutes in absolute alcohol, to which is added a little fuchsine or methyl blue as the case may be, and differentiation is effected with oil of cloves containing fluoresceine. They are thereupon cleared in etherial oil, the oil is extracted with xylol, and they are mounted in Canada balsam. Sections stained in methyl blue are transferred from the fluoresceine and oil of cloves to eosine and oil of cloves before being brought into etherial oil.

Kühne's dry method.—A one per cent. solution of ammonium carbonate is mixed with a concentrated aqueous solution of methyl blue, and this is allowed to act on the sections for ten to fifteen minutes. They are then washed in water, decolorised in an aqueous solution of hydrochloric acid, again washed in water, dried upon slides, cleared in xylol, and mounted in Canada balsam.

Weigert's iodine method.—By Weigert's method the sections are stained in aniline water gentian violet, rinsed in a solution of common salt, laid upon slides, and dried, and solution of iodine is dropped on them. After they have been again dried, aniline oil is poured over them and renewed several times. It is then removed with xylol, and the sections are mounted in Canada balsam.

A combination of Weigert's with Kühne's violet method (see p. 88) consists in staining the sections in a concentrated aqueous solution of violet to which some hydrochloric acid has been added (1 drop to 50 grms. water). After staining, the sections are rinsed in water, decolorised in iodine and potassium iodide solution, transferred to absolute alcohol, treated with aniline oil and with xylol, and preserved in Canada balsam.

Unna's borax methyl blue method.—A process which is particularly to be recommended for tubercle and lepra bacilli is the treatment of the sections for 5 minutes with aqueous *borax methyl blue*, from which they are transferred for 5 minutes more to a 5 per cent. solution of potassium iodide to which a crystal of iodine has been added. Rinsing in alcohol follows until a blue cloud forms, and then differentiation in creosote, lasting from a few seconds to half an hour, according to the intensity of the staining. The sections are afterwards transferred to rectified oil of turpentine, in which the bluish colour immediately changes to red or brown, and are put up in a solution of colophonium in oil of turpentine.

Unna's method of demonstrating the organisms of the skin. —Unna has devised several methods for staining micro-organisms in furuncles and abscesses of the skin, which can also be used to show the micro-organisms of pus. In all these methods the sections are previously stained in carmine and treated for two minutes with borax methyl blue (1 part each of borax and methyl blue in 100 of water), after which [in the *first method*] they are rinsed in water and placed for a few seconds in a 1 per cent. aqueous solution of arsenic acid, then in alcohol, bergamot oil, and balsam.

According to a *second method* the sections, after a slight preliminary staining with carmine and methyl blue, are brought for five to ten seconds into a 20 per cent. solution of ferrous sulphate, then into alcohol so long as any colour comes away, then for some seconds into a 1 per cent. solution of potassium binoxalate, and finally direct into absolute alcohol, oil of bergamot, and balsam.

In the *soap method* the previously stained sections are immersed in alcohol to which a few drops of *Spiritus Saponatus Kalinus* [1] have been added, and then into pure alcohol, bergamot oil, and balsam.

The *chromic method* consists in immersing the sections, after previous staining, in a 1 per cent. solution of potassium bichromate, washing them in water, and then transferring them for a considerable time into aniline oil, and finally from that into bergamot oil and balsam.

Noniewicz's method.—Noniewicz combined Löffler's and Unna's methods of staining in order to show the bacilli of glanders. The sections are transferred from alcohol to methyl blue for two to five minutes, rinsed in water and decolorised in a mixture of 75 parts of half per cent. acetic acid and 25 parts of half per cent. aqueous solution of tropæoline. Thin sections are only dipped quickly into the solution; thicker may remain in it for two to five seconds or longer. After being washed in water they are spread out upon a slide, dried in the air or over a flame, laid in xylol to clear, and mounted in Canada balsam.

Experiments on Living Animals

Transmission of micro-organisms to animals.—So far a series of methods of research has been described which are necessary for the diagnosis of bacteria; the observation of micro-organisms in the recent state, of their growth on different nutrient media, and of their behaviour in relation to staining materials forms, when taken together, the methods by which it is possible to demonstrate the micro-organisms

---

[1] [The *Spiritus Saponatus Kalinus* of the Austrian pharmacopœia consists of 200 parts of potash soap and 100 parts of spirit of lavender, prepared from lavender flowers by maceration and distillation.]—Tr.

in the tissues and fluids of the human body, as well as external to it. It still remains for us to give a brief account of those methods which are employed to ascertain the special significance of the different micro-organisms for the human body, that is to say, to recognise by means of experiment their *pathogenic* powers.

Amongst micro-organisms a distinction is drawn, as we have learnt, between those which exercise a specific injurious influence upon the bodies of men and animals, and those which do not possess this property, although they may perhaps occasion disturbances of various kinds by their numbers; the former being known as *parasites*, the latter as *saprophytes*.

In order, then, to investigate micro-organisms with reference to their power of causing disease, experiments must be made by transmitting them to animals, for which purpose monkeys, dogs, cats, hedgehogs, rabbits, guinea-pigs, white mice, rats, marmots, poultry, pigeons, or even frogs (kept at abnormally high temperatures) are used. [To prove that a particular micro-organism is the specific cause of a given disease it should be shown—[1]

1st. That its presence can be detected with the microscope *in all cases* of that disease.

2nd. That it is *never* found in any other disease.

3rd. That when isolated and cultivated through many generations a culture inoculated on a susceptible animal *invariably* produces a disease identical with that in the animal from which the virus was taken, and

4th. That the same bacteria are found to be present in the animal so inoculated.]

Transmission can easily be effected on the cutaneous surface, or on the mucous membranes of readily accessible

---

[1] [See on this subject Günther's *Einführ. in das Stud. der Bakteriol.*, pp. 139 *et seq.*, 2nd ed.]—Tr.

cavities, an experiment made in the latter way being often, indeed, attended by better results than follow transmission into small wounds of the skin ; but care must be taken that it does not become possible for the animals to remove the micro-organisms which have been introduced. Whether infection can take place through the epithelial structures of the skin, if unbroken, has not yet been finally decided.

Infection by the air passages.—Entrance can readily be effected through the respiratory tract; indeed, infection seems to be able to gain admission with particular ease by the internal surface of the lungs, especially sore ; the degree of moisture all over the surface assists by fixing the micro-organisms and enabling them to develop. For artificial infection by the respiratory passages a *spray apparatus* is used, by means of which the micro-organisms, suspended in bouillon, reach their destination in the form of a fine shower; but it is not easy to prevent the simultaneous occurrence of a second infection, since during the process the infectious matter may reach the intestinal canal by being swallowed, or may be deposited on the skin. To render this less easy of occurrence the excessively fine mist must be conducted by means of a tube into a closed chest in which the animal to be experimented on has been placed, so that it can thus freely breathe in the micro-organisms suspended in the air of the interior space.

Infection by the digestive canal.—Infection is communicated through the intestinal tract either in the food or directly by means of an œsophageal bougie, or the micro-organisms may be introduced by establishing a gastric or intestinal fistula. The best mode is, however, to hollow out pieces of potato, fill them with the bacterial culture, and push them so far back into the animal's pharynx that they must be swallowed. Fluid infecting material is administered to animals by means of œsophageal bougies introduced into the gullet—in the case of rabbits, through the gap between their teeth, in that of guinea-pigs, through a small perforated gag clamped between the incisors—the bougie used being a soft elastic catheter. When it is wished to infect an animal artificially the micro-organisms must be introduced into the intestine, as the acid gastric juice frequently impairs their vitality.

Nicati and Rietsch in their experiments on *cholera* injected the infecting liquid directly into the duodenum, which they had laid bare by a laparotomy performed with the strictest antiseptic precautions. Koch recommended the following mode of procedure for the purpose of excluding the injurious effect of the gastric juice on the micro-organisms:—A wooden gag perforated in the centre having been introduced into the mouth of the animal, a sound is inserted through it, and 5 c.cm. of a saturated solution of sodium carbonate is injected to neutralise the acid gastric juice. One grm. of tincture of opium for every 200 grms. of body-weight is then injected subcutaneously, in order to keep the animal in a state of narcosis, after which cholera-bacilli suspended in bouillon are injected by means of an œsophageal tube, and the experiment of introducing infection by the intestinal canal is complete.

Subcutaneous infection.—Inoculation can also be performed subcutaneously by introducing the infecting matter beneath the skin with a Koch's syringe. In the case of small animals, such as white mice, the hair of the back in the neighbourhood of the tail is carefully removed, a minute incision is made into the skin with disinfected instruments (forceps and scissors), and the infecting matter introduced subcutaneously with the help of a sterilised platinum loop.

Experiments of transmission into the peritoneal and pleural cavities, or into the organs themselves, are conducted after a similar fashion.

Intravenous infection.—This is most conveniently done into one of the superficial veins of the neck, or by puncture of an aural vessel.

Infection into the anterior chamber of the eye.—One of the most elegant modes of inoculation is the introduction of micro-organisms into the anterior chamber of the eye. This is done by opening the chamber with a lancet entered at the junction of the cornea and sclerotic, and introducing the infecting material through the wound so made. The aqueous humour which flows away is soon restored after cicatrisation of the wound, while the multiplication and visible peculiarities of the micro-organisms can be observed through the transparent cornea.

CHAPTER V

THE BACTERIOLOGICAL ANALYSIS OF AIR

Micro-organisms in the air.—Floating in the air are particles of dust consisting of organic substances, amongst which are also to be included, as a rule, dried-up colonies of micro-organisms. Such may either sink downwards of themselves under the influence of gravity, and so be caught, or they may be obtained by calling in the aid of currents of air, but in all cases they must be transmitted to a suitable nutrient medium before they can develop. As a rule we find in the air moulds, yeasts, and the spores of bacteria. On the open sea, far out from shore, the number of micro-organisms is considerably smaller, and in like manner the air on high mountains is almost entirely free from germs, or at least there are but few, whereas on the plains 100 to 500 germs capable of living have been counted in each cubic centimeter. The air of dwelling-rooms contains them in considerable numbers only when they have been whirled up from between the flooring and from the coatings of the walls, and this detachment of bacteria by draughts of air can only take place when the surfaces are dry.

Simple methods of examining air.—The simplest way of examining air consists in letting a plate prepared with agar or gelatine stand in any locality for a definite time, and afterwards placing it in a moist chamber, when colonies of micro-organisms will form in a few days. Agar plates

POUCHET'S METHOD OF ANALYSIS

may also be placed in the incubator, in order to observe the micro-organisms which develop at a higher temperature.

The method can be simplified by pouring the gelatine into capsules, which, after catching the germs from the air, are closed and kept. Such capsules may be exposed in a glass vessel of cylindrical form, the volume of air in which is known; after a fixed time the process can be stopped, and the capsules with the gelatine set aside for the organisms to develop. Knowing the volume and the time of exposure,

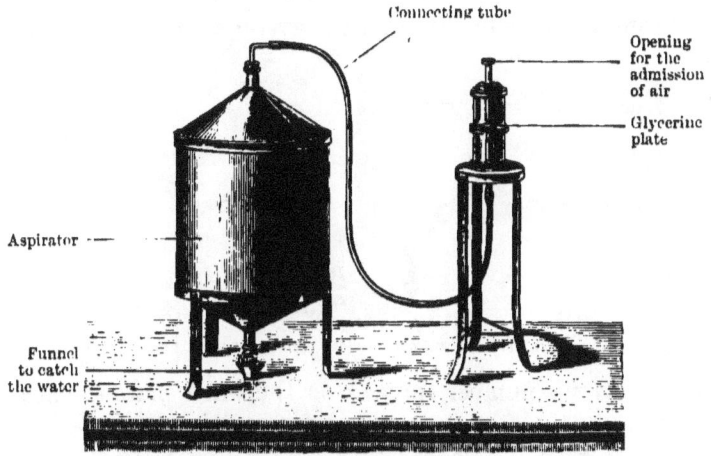

FIG. 28.—POUCHET'S AËROSCOPE.

it is possible to gain an approximate idea of the number of micro-organisms contained in the air.

Pouchet's method.—Pouchet employed for the examination of the dust of the air an *aëroscope* consisting of a glass cylinder, capable of being closed air-tight by means of a screw and clamps; it is placed vertically upon a stand and perforated above and below. In the upper aperture is a glass tube with a very narrow exit, the lower one communicates with an aspirator through an indiarubber pipe, and in the centre of the cylinder is a little table supporting a small glass plate, which is smeared with glycerine. The aspirator being put in action, air streams in through the upper aperture and deposits the greater part of the dust it contains upon the glycerine, and the preparation is removed from the cylinder and examined as soon as sufficient air has been drawn through. The dust is distributed as evenly as possible through the glycerine by stirring with a sterilised steel needle, and the glass plate is covered with a second and brought under the microscope. To calculate the amount of dust in a litre of air, the particles in several microscopic fields are counted, so as to ascertain the average number in each;

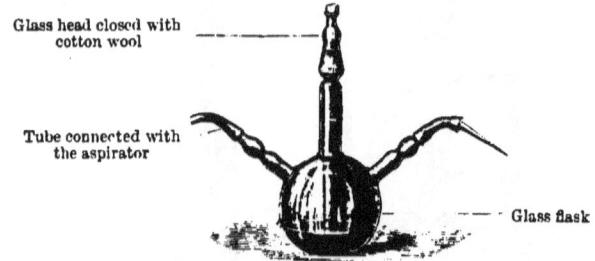

FIG. 29.—MIQUEL'S APPARATUS FOR EXAMINING AIR.

from this the number spread over the whole plate is calculated, and thence the amount contained in a litre.

Instead of the glycerine plate one of gelatine or agar may be laid on the little table, and an attempt thus made to isolate the micro-organisms (fig. 28).

Miquel's method.—Miquel constructed a flask with two lateral tubes (fig. 29) and another fitting by a ground joint into the aperture at the top, and supporting a cap or head of glass closed with a cotton-wool plug. One of the lateral tubes is connected with an aspirator, the other (by means of a piece of rubber piping) with a narrow glass tube sealed at one end. The flask is filled with 30 to 40 c.cm. water, and sterilised in the steam current; the glass cap is then taken off and a given volume of air aspirated through, after which the cap is again put on, and, by blowing air through the lateral tube which was connected with the aspirator, the fluid is driven up into the vertical one, so as to wash it out. Finally the point of the glass tube on the opposite side is broken off, and the fluid contained in the flask distributed into tubes of bouillon.

Emmerich's method.—In the apparatus devised by Emmerich for bacteriological research of this nature, the air is drawn slowly through a coiled tube filled with nutrient bouillon, and the germs are in this manner retained (fig 30).

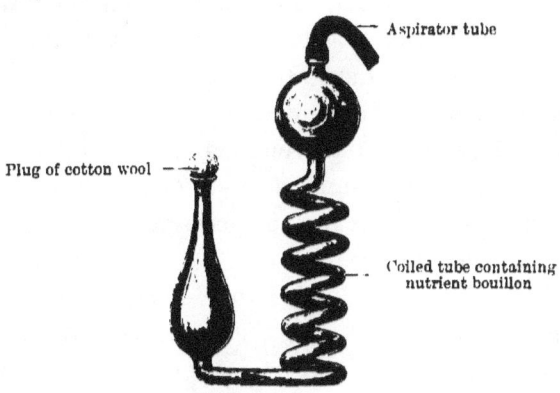

FIG. 30.—EMMERICH'S APPARATUS FOR EXAMINING AIR.

Welz's method.—Two small flasks, one as a receiver, and the other as a control flask, are prepared with 20 c.cm. each of a neutral liquid composed of equal parts of glycerine, bouillon, and water, and are connected together by means of a glass tube bent twice at right angles, the longer limb of which reaches to the bottom of the control flask, the shorter to just below the stopper of the receiving flask. Two large flasks connected by means of a rubber tube are used for aspirating, one being filled with water and united to the controlling flask. The other, which is empty, stands at a lower level than that containing the water, so that this may be able to flow into it; and in this way a volume of air, corresponding to the quantity of water used, is aspirated into the receiving bottle. For the purpose of regulating the flow, two little glass tubes, drawn out to fine points, are fixed in the india-rubber tube which connects the two aspirating bottles. Cultivation is effected by conveying 1 c.cm. of the fluid in the receiving flask (after it has been thoroughly mixed) by means of a sterile pipette into 10 c.cm. gelatine, and pouring this out on plates.

Hesse's method.—In this method the air is caused to pass by means of a small slowly-acting aspirator through a disinfected tube, the walls of which are coated with gelatine after the manner of Esmarch's roll cultures. This tube is 70 cm. long, and has a diameter of 3 to 4 cm.; it is placed horizontally, and covered at one end with a tightly-stretched rubber cap having a round piece cut out of the centre, and over which a second cap, not perforated, can be drawn; while the other end is closed with a caoutchouc cork, bored, and fitted with a small glass tube about 1 cm. wide and 10 cm. long, connected with the aspirator. While the air is being aspirated, the unperforated cap must be removed. Two bottles are used by way of aspirator, as in Welz's method, one filled with water, the other empty (fig. 31). The bacteria develop chiefly in the fore part of the tube, while the spores of moulds, being isolated and therefore lighter, are carried further and develop further on in the interior. When air is examined which presumably contains but few germs—for example, air out of doors during a calm—10 to 20 litres are drawn through, but if it is probable that large numbers are present only 1 to 5 litres are aspirated. The process is concluded by replacing the unperforated rubber cap. In a few days the gelatine is seen to be covered with colonies which can be distinguished from one another by their form, their colour, and their action on the gelatine (liquefaction). The germs may then be isolated by further transference to culture plates, and submitted to microscopic examination.

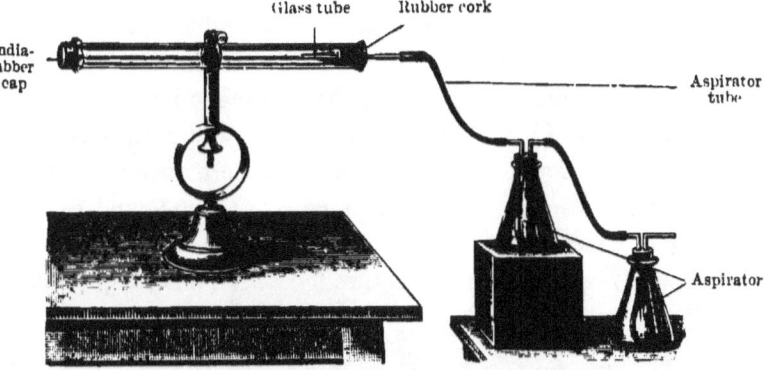

FIG. 31.—HESSE'S APPARATUS FOR EXAMINING AIR.

Method of Strauss and Würz.—Air is drawn into a glass vessel full of liquid gelatine by means of a tube affixed to

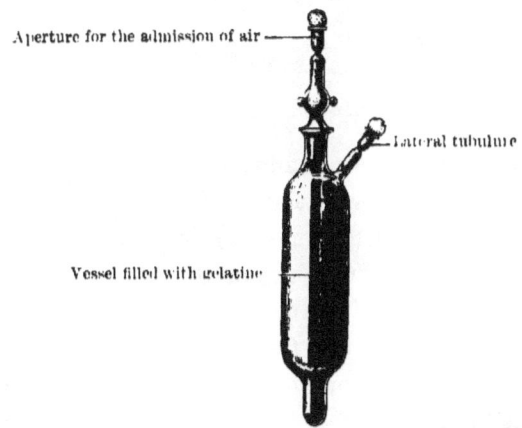

FIG. 32.—AIR-TESTING APPARATUS OF STRAUSS AND WÜRZ.

the side and connected with an aspirator (fig. 32). A large volume of air, 100 to 200 litres, is thus tested, and when the aspiration is concluded the gelatine is poured out on plates, or a roll-culture is made after Esmarch's method in the vessel itself.

Petri's Method.—A sand-filter is prepared and carefully sterilised. This consists of a tube 8 or 9 cm. long and 1·5 cm. in diameter, into which sand is introduced after one end has been closed with wire netting. When the layer of sand has reached a depth of 3 cm. another wire netting is laid on it, [another layer of sand introduced, and a third netting last of all], so that the tube is now provided with two sand-filters kept together by wire gauze. Quartz-sand, each grain of which is $\frac{1}{4}$ to $\frac{1}{2}$ mm. in size, is the best. About 50 to 100 litres of air are drawn through with a water aspirator

FIG. 33.—PETRI'S SAND-FILTERING APPARATUS.

at the rate of some 10 litres per minute, the quantity being determined by means of a gas meter. When aspiration of the air is concluded, each sand-filter is partitioned out separately into several glass capsules prepared with nutrient gelatine or some other solid culture medium. The second layer of the filter should be free from germs.

Miquel uses powdered sodium sulphate to absorb the microbes instead of sand.

Tyndall's method, &c.—Sterilised cotton wool is used for absorbing the micro-organisms, instead of air-filters consisting of substances in the form of powder, and is then transferred to gelatine, and plate cultures made therefrom.

Percy Frankland uses glass wool instead of cotton.

With the aid of these methods, *moulds, yeasts, micrococci, bacilli,* and *spirilla* are found, all of which are contained in greater or less quantity in the air, though their distribution is not the same in all parts of the earth's surface, nor at all times, either as regards quantity or quality. For example, the author found that the *Micrococcus prodigiosus* grew in abundance on a paste medium in the Alps (Hollenegg, Styria) in the month of September, 1891, whereas in the months of July and August in the same year no perceptible trace could be found.

Penicillium Glaucum.—The *Penicillium glaucum*, or pencil fungus, grows in the form of locks of cotton-wool, and during sporulation forms a green fur of a peculiar musty odour. Its *mycelium* consists of horizontally-arranged, straight, or slightly undulating jointed filaments, from which the spore-bearing *hyphæ* (*Fruchthyphen*) stand vertically up, dividing at their upper ends into forks (*basidia*) from which fine processes branch off (*sterigmata*) in the shape of a hair pencil, and are segmented at their ends into rows of fine globular bodies (spores or *conidia*), which in the mass give the fur its green colour (see fig. 4). Sterilised bread-pap is particularly well adapted for the growth of the pencil fungus, which forms a fur upon it, white at first, but afterwards taking on a fine green colour; but besides this it grows in all sorts of places where as a rule only mould can develop. Gelatine is liquefied by it. The growth of mycelium takes place very well according to Wiesner at a temperature of 26° C.; sporulation progresses best at 22° C.

The fungi appear on plate cultures first as threads diverging from a point, and do not form sharply-defined dark-coloured colonies upon the gelatine, but radiate out over a considerable extent of surface. The spore-bearing hyphæ which rise free above the level of the gelatine are put in motion by currents of air (such as lightly blowing upon the plate culture), and when this occurs the shedding of the spores can be readily observed. The earliest formation of spores occurs in the centre of the colonies, and is indicated by a green coloration.

Löffler's methyl blue stains the filaments of mycelium and the hyphæ, the spores on the other hand remaining unstained. Moulds cannot easily be moistened with water, as their surface has no affinity for it, owing to the presence of a thin coating of fat; hence the first proceeding is to treat the unstained moulds with alcohol to which a little ammonia has been added, after which they are examined in glycerine and water or plain glycerine. For making permanent preparations, glycerine or glycerine jelly is suitable, the cover-glass being cemented with asphalt lac dissolved in turpentine over a water-bath.

Hansen recommends the addition of 0·1 to 0·2 per cent. of hydrochloric acid when growing moulds upon gelatine, in order to keep away bacteria.

Brown mould.—The fur formed by the *brown mould* described by Hesse is brownish yellow, and is further distinguished from penicilium by its closely felted mycelium, the hyphæ being scanty, ramified, and segmented. Gelatine is very rapidly liquefied, and in thrust-cultures becomes softened with the mycelium of the fungus into a brown, viscid, stringy mass. Growth takes place best at 15° to 20°C. According to Trelease the mould is identical with an alga, the *Cladothrix dichotoma*, very frequently found in water.

Yeast.—This micro-organism consists of cells and masses of cells of which the individual elements possess an oval figure and multiply by gemmation. They have a thin limiting membrane and a granular protoplasm containing vacuoles (see fig. 6). They are obtained from the air upon gelatine or agar plates, or upon sterilised potatoes, and grow on the first-named in the form of round colonies raised into knobs of a drop-like appearance, which do not liquefy the medium.

Pink yeast is distinguished by the rose-pink colour of the mass, and shows on gelatine-plates small round, rather coarsely-granulated, rose-coloured colonies. In thrust cultures there appears after eight days a coating with a dull surface like a drop of wax, which slowly increases in circumference and shows raised edges, while a row of little dots forms along the track of the thrust. The gelatine is not liquefied. On agar there grows an irregular thin slimy coating of a pink colour, and on potato a deposit of a beautiful rose-red tint.

Black yeast and **white yeast** only differ in the colour of the coatings formed by them.

Yeast grows at the temperature of an ordinary room. When stained with aniline colours the cells shrivel somewhat, and do not show the fine figures seen if they are examined in the unstained condition. They can easily be distinguished from cocci by their remarkable size (1·5 μ to 3 μ long, 2 μ broad).

Micrococcus radiatus.—The *Micrococcus radiatus*, isolated by Flügge, forms small cocci in short chains or clumps. On gelatine plates little yellowish-brown colonies first appear, from which outgrowths push forth in a radial direction, and in thrust-cultures rays are seen running out horizontally from the centre of the track, so that it acquires an almost feathery appearance. The gelatine is slowly liquefied. Colonies of a yellowish colour form on potatoes.

Micrococcus versicolor.—This micro-organism, which was also found by Flügge, is distinguished by the mother-of-pearl gloss seen on its colonies. It forms minute clumps of cocci, which appear on the very first day as round white points on the gelatine plate, and these after several days become yellowish-brown and in certain positions of the plate shimmer like mother-of-pearl. The latter appearance is also seen on the surface of a thrust-culture as well as on superficial cultures upon agar. The shimmer sometimes strikes into a yellowish-green. Gelatine is not liquefied.

Micrococcus cinabareus.—The microbe so named by Flügge, and which is noticeable on account of its cinnabar colour, forms diplococci. On gelatine plates the colonies do not appear for four to six days, and are reddish-brown; but the colour gradually changes to vermilion. A thrust-culture also shows the colour on the surface in the form of a red knob, from which a white stripe extends into the gelatine, which is not liquefied.

Micrococcus flavus tardigradus of Flügge appears in large single elements, and forms colonies which become yellow in six to eight days. In the thrust-canal isolated and disconnected yellowish balls develop. It is distinguished by its yellowish colour, and does not liquefy gelatine.

Micrococcus candicans (Flügge) often appears as a contamination on gelatine plates, forming white colonies which are darker in the centre and lighter towards the margin. Thrust-cultures are nail-shaped with a nodular elevation. Gelatine is not liquefied.

Micrococcus viticulosus.—This micro-organism, described by Katz, consists of oval elements, and is particularly characterised by the tendril-like shapes of its colonies. These appear both on and below the surface of the gelatine plates, and send out lateral outgrowths in the form of fine processes resembling tendrils. In thrust-cultures it grows along the track of the puncture, from which a delicate network of fibres likewise runs into the substance of the gelatine without liquefying it.

Micrococcus ureæ.—This is a micro-organism which does not liquefy gelatine, and which, although it occurs in the air, can also be obtained from decomposed ammoniacal urine. It was described by Pasteur and Van Tieghem.

The cocci are for the most part arranged in pairs, and when they have grown for a considerable time upon gelatine, show a rather large-sized colony which is raised above the surface like a drop of solidified stearine, and diffuses a smell like that of paste. It sets up fermentative processes in urine, owing to its property of converting urea into ammonium carbonate. Growth takes place at room-temperature, or best at 30° C., but it does not lose the power of development even at temperatures below zero (fig. 34).

FIG. 34.—MICROCOCCUS UREÆ. (After Jaksch.)

Micrococcus roseus.—This is a micrococcus occurring in gonorrhœal disease of mucous membranes, and was found in the air by Bumm. The cocci are immotile, and are arranged in pairs, each half of the diploccocus so formed being hemispherical, and separated from the other by a fissure. The small colonies formed by it on the gelatine plate are rose-red and raised above the surface. Thrust-cultures show liquefaction of the gelatine, but only after a considerable time, and a rose-red colour then forms at the bottom of the needle-track.

Diplococcus citreus conglomeratus.—This micro-organism was found by Bumm in the dust of the atmosphere, with which it probably enters the human organism. Its elements are not motile, and are sometimes arranged in pairs, at other times in fours. They form small oblong colonies on the gelatine plate, which soon become fissured, and possess a lemon-yellow colour; the thrust-culture slowly liquefies the gelatine, a yellow mass being found in the deeper part. Surface cultures on agar exhibit a lemon-yellow coating, which later becomes brownish.

Micrococcus flavus liquefaciens and Micrococcus desidens. —Both have been described by Flügge. The former has larger, the latter smaller elements frequently arranged as diplococci, and both form yellowish-coloured collections on discs of potato. On gelatine plates small round yellowish colonies occur, which begin to liquefy the gelatine in from one to two days. Thrust-cultures liquefy in a few days, earlier in the case of *Micrococcus liquefaciens* than with *Micrococcus desidens*, and when the elements have sunk to the bottom of the funnel-shaped fluid area, a slight yellow coloration forms below.

Sarcina alba.—The *Sarcina alba* grows slowly on gelatine plates in little round white colonies, and in the same manner along the track of the thrust in the test-tube, forming in the latter a small white head on the surface. It also grows very slowly on potatoes, in the form of a whitish-yellow deposit round the site of inoculation. Gelatine is very slowly and only very slightly liquefied.

Sarcina candida.—The microbe of this name, found by Reinke in breweries, shows shining white colonies on gelatine, which later become yellowish and very soon liquefy. On agar there appears a white deposit with smooth edges.

Sarcina aurantiaca.—Small smooth-edged colonies appear on the gelatine plate, having a dotted granular aspect when seen under a low power, and an orange-yellow colour. Thrust-cultures liquefy slowly along the entire track, and excrete an orange-yellow pigment at the surface; but when they have stood for a longer time, the principal mass sinks to the bottom and the superficial part of the medium becomes clear. Agar cultures also show a fine golden-yellow glossy coat. The sarcina grows slowly on potato. Gelatine is but little liquefied. Sulphuric acid turns the golden-yellow pigment bluish-green, and caustic potash, red.

Sarcina rosea.—This micro-organism, which was discovered by Schröter, grows very rapidly on gelatine, slowly on agar, forming minute cartilaginoid clumps; while a vigorous, intensely red deposit forms on potato. Multiplication takes place in broth with extraordinary rapidity, and with development of a red sediment. Gelatine is very speedily liquefied. The red pigment exhibits the same chemical reactions as the colouring matter of *Sarcina aurantiaca*.

Sarcina lutea.—The sarcina of this name, described by Schröter, grows very slowly on the gelatine plate, forming small round colonies. A scanty coating appears diffused over the surface, and advances into the deeper parts over a narrow area in the form of yellow granules. A thickish deposit of a fine yellow colour appears upon agar. Cultures on potato are sulphur-yellow, and confined to the place of inoculation. Gelatine is sometimes liquefied slowly.

Staphylococci.—The *Staphylococcus pyogenes* was fully described by Rosenbach, by Ogston, and by Passet. Its distinctive characteristic is its power of causing suppuration, and it may consequently be described as a specific pus-coccus, being constantly found in suppurative processes. There are distinguished, according to colour, a *Staphylococcus pyogenes aureus*, a *Staphylococcus pyogenes albus*, and a *Stapylococcus pyogenes citreus*. It is but seldom in the analysis of air that a plate culture destitute of staphylococci is obtained.

According to E. Ullmann, these micro-organisms are found in considerably greater numbers in the air of rooms which are much used than in localities but little frequented by human beings. Ullmann also found them very widely diffused elsewhere in nature, not only in the air but in river-water and rain-water, though not in spring-water; also in ice, in the earth, and on walls.

They are small globular immotile cells, always tending to form closely-packed clusters, particularly in the interior of tissue (fig. 35). Those cells, however, which are not included in the clusters possess the power of moving with tolerable activity. The individual cells take up all the different aniline dyes, grow even at ordinary temperatures, though more energetically at degrees of heat approaching that of the human body, and, if added to sterilised milk, precipitate the caseine.

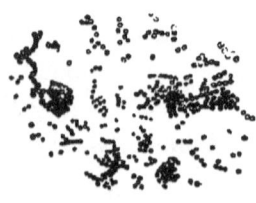

FIG. 35.—STAPHYLOCOC PYOGENES AUREUS.

The **Staphylococcus pyogenes aureus** grows very quickly on the gelatine plate at the temperature of an ordinary

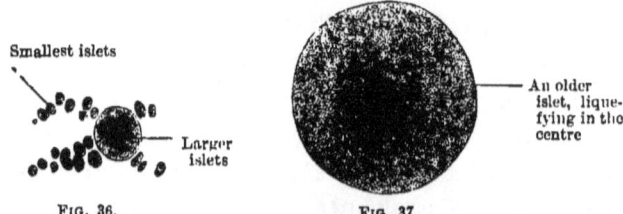

FIG. 36.   FIG. 37.
ISLETS OF STAPHYLOCOCCUS PYOGENES AUREUS ON A GELATINE PLATE.

room, so that even as early as the second day small punctiform colonies are to be seen, which are round and possess a sharply-defined circumference, and these soon approach the surface and liquefy. The liquefaction extends out at the periphery, and soon shows a yellowish colour in the centre (figs. 36 and 37). In the thrust-culture the gelatine begins to undergo liquefaction on the second or third day, and this gradually advances deeper along the thrust-canal, while, as the funnel-shaped liquid area enlarges, the cocci

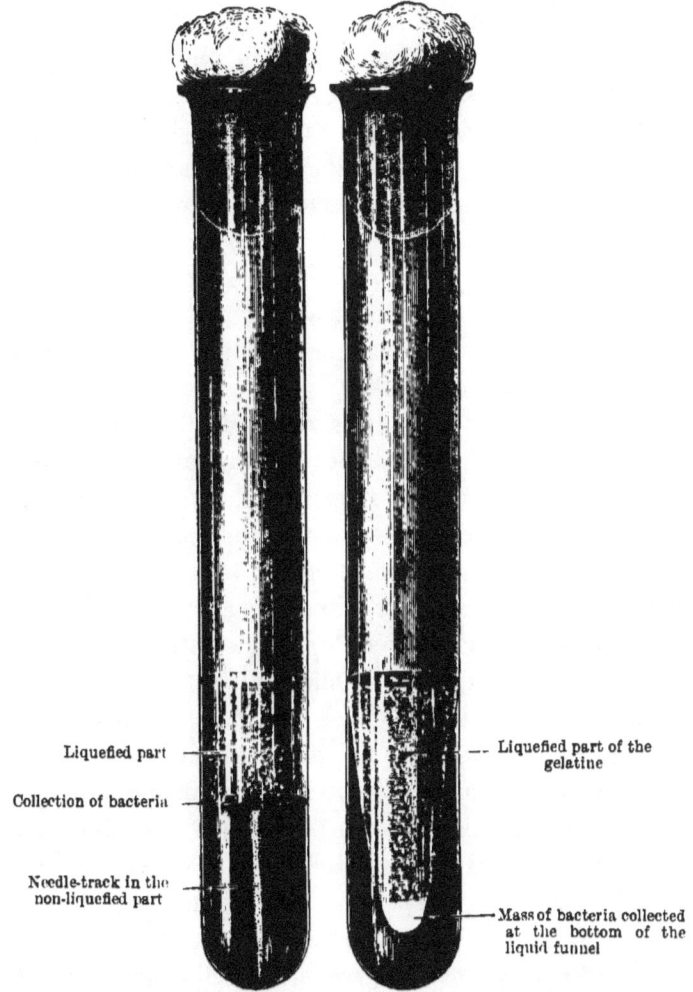

FIG. 38.—THRUST-CULTURES IN GELATINE OF STAPHYLOCOCCUS PYOGENES AUREUS; —TWO DIFFERENT FORMS OF GROWTH.

Liquefied part
Collection of bacteria
Needle-track in the non-liquefied part
Liquefied part of the gelatine
Mass of bacteria collected at the bottom of the liquid funnel

sink to the bottom and begin to take on colour (fig. 38). When, however, the culture is exposed, not to ordinary temperature, but to a higher degree of heat, although, indeed, the growth is more luxuriant, there is not so fine a formation of colour as at the former temperature. This is true particularly of surface cultures on agar, in which a thick column forms at first along the streak, and then gradually spreads out further so as to cover the surface of the agar with a complete coating of culture displaying the characteristic colour. On potatoes there occurs a deposit which is at first whitish but afterwards takes on a yellow or orange hue.

Staphylococci also grow excellently on serum and the white of plovers' or pigeons' eggs. All cultures of them very soon develop a strong smell of paste, which, as the age of the cultures advances, is modified to an odour resembling that of sour milk.

Successful infections have repeatedly been made with staphylococci. When brought upon the surface of wounds, they set up a progressive suppuration, while subcutaneous injections originate abscesses, and injections into the circulation cause inflammation of joints and abscesses in the kidneys and myocardium. According to Orth, Wyssokowitsch and Ribbert, they set up an ulcerative endocarditis on diseased or perforated cardiac valves. All these phenomena are dependent either on the occurrence of a mechanical derangement of the vascular areas by the micro-organisms, or on the development of metabolic products having a toxic action on the tissues. In addition to entering by wounds, the staphylococci can find their way into the cutaneous and subcutaneous tissue from the hair follicles and the ducts of the cutaneous glands.

The **Staphylococcus pyogenes albus** is distinguished from the last only by the absence of pigment; it appears to be less energetic in its action. **Staphylococcus pyogenes citreus** differs also in colour, and liquefies gelatine more slowly than either of the other two.

Leber isolated the active principle from the cultures in the form of a crystalline body, to which he gave the name of *phlogosin*, and which, if injected in small quantity, leads to suppuration without the presence of micro-organisms. Christmas obtained a pyogenic body from the cultures in the shape of a substance of the nature of a ferment, which set up suppuration when introduced into the anterior chamber of a rabbit's eye. E. Ullmann caused osteomyelitis by intravenous injection of dead cultures after a previous fracture.

Streptococci.—Emmerich and Hartmann succeeded in isolating a streptococcus from the air, which, when inoculated on rabbits, set up a typical erysipelas, and is therefore described as *Streptococcus erysipelatis*. A pure cultivation was first obtained by Fehleisen. Gelatine is not liquefied. On the plate small colonies appear in the substance of the gelatine on the third or fourth day, and gradually assume a brownish colour. In thrust cultures the superficial growth is very scanty, but along the needle-track very minute white globular colonies appear, forming a white stripe. Small round isolated colonies develop upon agar, resembling drops of dew. No growth takes place on potatoes.

According to Jordan, Fränkel, and Von Eiselsberg, it is identical with the *Streptococcus pyogenes* (see p. 201).

Bacillus subtilis.—This bacillus, also called the *hay bacillus*, was described by Ehrenberg, and is most easily obtained from an infusion of hay made by chopping up the hay, pouring water on it in a flask, and bringing it once to the boil. In this way all the other different micro-organisms are easily killed, the hay bacillus alone suffering no impairment of vitality. After two or three days a thick whitish pellicle forms on the surface, and consists of a pure culture of the *Bacillus subtilis*. The bacillus takes the form of very long, fine thin rods, possessing marked power of movement by means of flagella, and a disposition to unite into groups. Owing to their motility, the bacilli, or the threads formed by them, are seen to dart with an undulating motion across the field of the microscope. On the gelatine plate little white dots occur, which soon extend and liquefy the gelatine over a still wider surrounding area, while around the liquefied mass fibres of bacilli are moreover seen growing into the gelatine in the form of a halo. Thrust-cultures likewise show an energetic liquefaction (fig. 39), and as soon as the gelatine in the test-tube has become completely fluid a coating or pellicle forms on the surface. An extensive growth develops upon agar, and on potato there appears a creamy deposit, which in a few days takes the colour of wine. Serum and the coagulated albumen from plovers' and pigeons' eggs are liquefied, and on these also the superficial formation of membrane is very marked. According to Wyssokowitsch, if the spores are introduced into the circulation they expand into rods, and remain lying in the liver and spleen without exercising any influence on the organism. According to Vandervelde, the *Bacillus subtilis* sets up active fermentation of sugar.

Bacillus prodigiosus.—The *Bacillus prodigiosus*, which is especially remarkable on account of the development of a red pigment, falls from the air at certain times upon substances containing starch, on which it grows with tolerable rapidity. and it has thus given origin to the legends of showers of blood. The rods are so very short that their long diameter scarcely exceeds their breadth, and for this reason the bacillus was formerly classed with the micrococci. The individual rods are motile. On acid nutrient media, however, they expand, according to Kübler, into larger bacilli, which also possess the power of motion. They form spores. On gelatine plates they show even after ten or twelve hours small round granular colonies, which soon liquefy from the surface downwards and coalesce with one another if they lie closely, the diagnosis being established by the early appearance of a

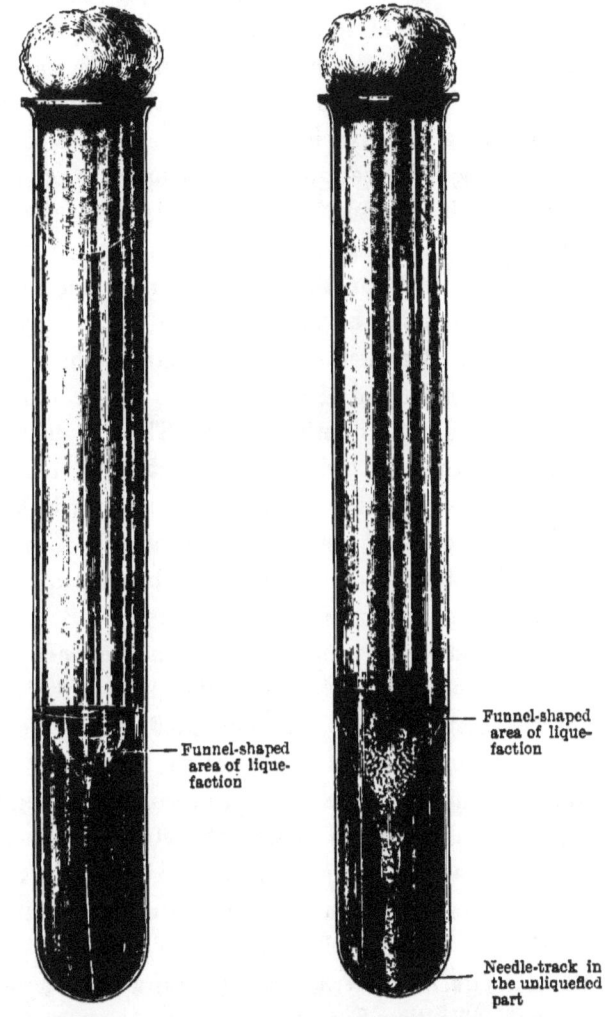

Fig. 39.—Thrust-Culture in Gelatine of Bacillus subtilis (third day).

Fig. 40.—Thrust-Culture in Gelatine of Bacillus prodigiosus (fourth day).

red colour. The thrust-culture liquefies from the surface down, and soon a funnel-shaped area of liquefaction is formed, upon the surface of which the pigmentation takes place, owing to contact with the air, and then sinks gradually downwards (fig. 40). A beautiful purple-red colour develops on the surface of streak-cultures on agar, but the finest growth takes place upon slices of potato or wafers, on which, moreover, it progresses very rapidly at the temperature of the room, being less luxuriant at higher degrees of heat. The bacillus liquefies serum, and soon appears on plovers' egg albumen with a beautiful rose-red colour, which extends only as far as the coagulated mass has become liquid. The coating shows a punctiform appearance under a low power. The spectrum of the pigment, which is readily soluble in water, alcohol, and ether, has three absorption bands, one towards the violet end of the spectrum at [Fraunhofer's line] D, one at E, and another at F. The red colour becomes brown in the air, owing to the action of ammonia, but recovers its raspberry-red colour if acetic acid be applied.

Wyssokowitsch, and quite recently E. Ullmann, have proved that dead cultures of this bacillus are capable of exciting suppuration, and Grawitz and De Bary found that its pathogenesis is connected with the pigment.

Potato bacillus.—Three varieties are distinguished: **Bacillus mesentericus fuscus** (Flügge), **Bacillus mesentericus ruber** (Globig), and **Bacillus mesentericus vulgatus**. They show short filaments which are often connected together into chains, and have the power of active movement. They liquefy gelatine very quickly during their growth, whether on the plate or in thrust-cultures, and form round colonies which soon become yellowish, and in the case of the brown bacillus (*Bacillus mesentericus fuscus*) assume a dark brown colour. The liquefied gelatine, which swarms with bacilli, also darkens (fig. 41). Upon discs of potato they grow very luxuriantly, and soon spread from the upper to the lower surface. The *Bacillus mesentericus ruber* shows at a higher

BACILLLUS MESENTERICUS

temperature (of about 37° C.) a reddish-yellow or rose-red colour. The individuals of all the varieties included under the name of *potato bacillus* adhere together and form an extensive wrinkled membrane, which can easily be detached from the slice of potato. The *Bacillus mesentericus vulgatus* has the property of curdling milk, as rennet does, and rendering it stringy, the substance to which it owes its viscidity being probably metamorphosed cellulose. It displays upon the whole the same behaviour towards gelatine and agar that the two other potato bacilli do, but whereas the cultures of *Bacillus mesentericus fuscus* have a yellowish colour, and those of the *ruber* variety a reddish, the membrane on the

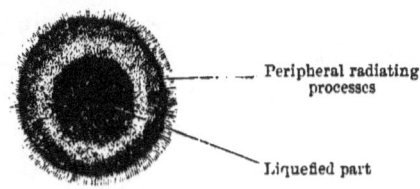

FIG. 41.—ISLET OF BACILLUS MESENTERICUS VULGATUS ON A GELATINE PLATE.

potato shows no pigmentation at all in the case of *Bacillus mesentericus vulgatus*.

The potato bacillus develops with particular readiness on pieces of potato which are not completely sterilised, often destroying the cultures of other micro-organisms.

Bacillus liodermos.—Flügge found very widely distributed in the atmosphere, and often as a guest upon our nutrient materials, short, exceedingly motile rods, the growth of which on gelatine causes it to liquefy with great rapidity, a white pellicle floating on the surface. In thrust-cultures dirty grey flakes swim about in the fluid mass. A smooth, glossy coat resembling thin mucilage develops on potato, changing as the spores form into a thick and much-wrinkled membrane. The mucilaginous mass is soluble in water. The *Bacillus liodermos* grows with especial luxuriance in milk.

Bacillus melochloros.—The *Bacillus melochloros* was originally discovered in the author's Institute by F. Winkler and Von Schrötter, in the caterpillars' excreta found in worm-eaten apples. It is at times a constant inhabitant of the air of the author's laboratory, and often appears as a guest upon cultures of other micro-organisms; and it is possibly identical with the *Bacillus butyri fluorescens*, found by Lafar in butter. It consists of slender, fairly long rods, with smoothly rounded ends and actively motile, and is distinguished by its unusually rapid growth, so rapid that even in four hours there appear on the plate greyish-white colonies, in which darker and more closely-packed masses are to be seen; while as early as the second day the gelatine is liquefied with development of a greenish-yellow colour. In thrust-cultures also, on the second day, an hourglass-shaped depression shows itself, around which there is very rapid liquefaction (fig. 42). The speedy growth and greenish-yellow colour are also seen in superficial cultures on agar, the surface of which very soon becomes overspread with a thick yellowish coating, while all the rest of the medium acquires a green tinge. On plovers' egg albumen it grows with a splendid emerald green colour, and on potato it forms a dirty reddish-yellow layer. The pigment developed by the *Bacillus melochloros* is very readily soluble in water, but not at all in alcohol or chloroform. It is destroyed by acids, but restored again by alkalies. Older cultures acquire an exceedingly unpleasant odour. When the pure culture is injected into the veins or peritoneal cavity of rabbits the animals perish in a week at furthest.

Bacillus multipediculosus, which was discovered by Flügge, shows small thin immotile rods. The colonies on a gela-

BACILLUS MULTIPEDICULOSUS 119

tine plate appear under a low power as circular, sharply-defined discs with radiating processes, resembling insects

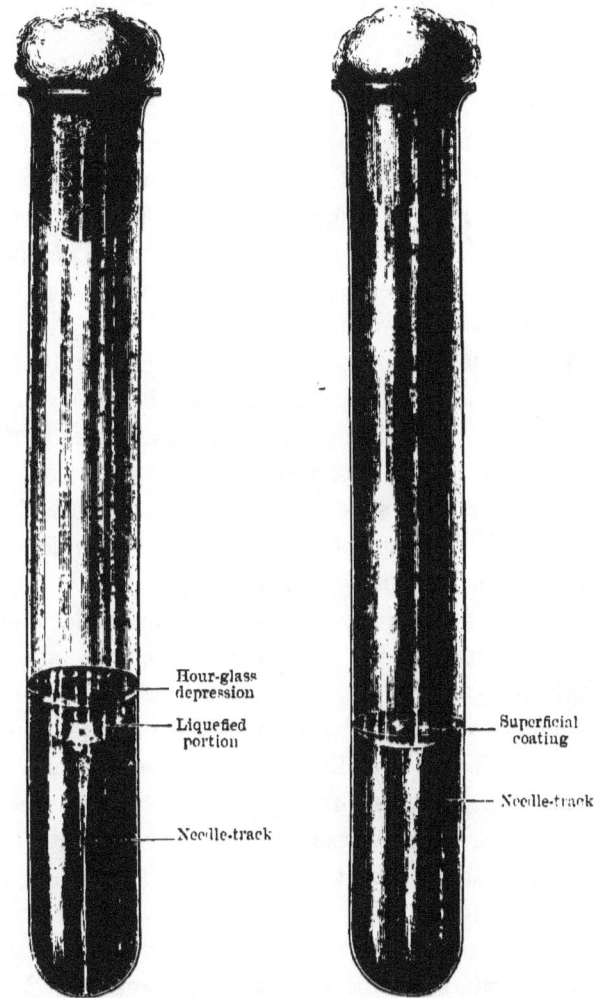

FIG. 42.—THRUST-CULTURE IN GELATINE OF BACILLUS MELOCHLOROS (SECOND DAY).

FIG. 43.—THRUST-CULTURE IN GELATINE OF EMMERICH'S BACILLUS (*Bac. Neapolitanus*).

with numerous radially arranged feet. In thrust-cultures also the processes appear extending from the needle-track in all possible directions, and these peculiar projections have procured for this micro-organism its name of '*multi-pediculosus*.' Gelatine is not liquefied.

Bacillus neapolitanus.—This microbe was first discovered by Emmerich in the blood and in evacuations from the corpses of cholera patients in Naples, and it was subsequently ascertained to be present in normal fæces. Pathogenic powers were ascribed to it, because a disease resembling the cholera in human beings develops after the introduction of considerable quantities of it into the bodies of guinea-pigs, dogs, cats, and monkeys; the introduction may be effected subcutaneously, or into the abdominal cavity or the lungs. Microscopic examination demonstrates the bacilli in all the organs. There have, however, been objections made by Weiser to ascribing pathogenic properties to it, and he has shown that it is present in the air also.

The bacillus appears as a short rodlet with rounded ends and destitute of motile power, which forms on the gelatine plate colonies resembling porcelain and lying at a greater or less depth, of which the superficial ones spread as a coating over the surface of the gelatine, and the deep have a figure like that of a whetstone. In thrust-cultures the more vigorous growth takes place on the surface (fig. 43). Gelatine is not liquefied, but loses its alkalinity, which causes a clouding of the transparent jelly and a simultaneous separation out of crystals of salt. If tincture of litmus is added to the gelatine the blue colour disappears and becomes changed to a red. A dirty white mass forms on agar and potatoes. With regard to staining processes, it is a special characteristic of this micro-organism that it does not colour by Gram's method. Its resistance to external influences is so great that it retains its vitality after being frozen for twelve days and then thawed again.

Atmospheric spirilla.—The spirilla occurring in the air have been described by Weibel. They usually generate yellow pigment, according to the degree of intensity of which there have been distinguished a **Vibrio aureus** with a colour varying from golden to orange-yellow, a **Vibrio flavus** of an ochre-yellow tint, and a yellowish-green **Vibrio flavescens.** The individual spirilla frequently appear remarkably thin, generally S-shaped, and without power of automatic movement. Islets of an oval or whetstone form, or sometimes circular, develop on the gelatine plate; they are sharp-edged and granular, and generate pigment in a few days. There is no liquefaction. Thrust-cultures in gelatine and superficial cultures on agar display also a copious development of colour, which takes place only on the surface of the former; while on discs of potato there appears a luxuriant pap-like deposit of a very pronounced tint. Spirilla are decolorised by Gram's method.

CHAPTER VI

THE BACTERIOLOGICAL ANALYSIS OF WATER

Micro-organisms of water.—Water, both in its liquid and solid state, almost always contains micro-organisms, although in variable quantity, and these have been named *water bacteria* by Percy Frankland. They are for the most part bacilli—in general such as do not liquefy gelatine—and they do not grow at the higher degrees of temperature. Some of them have the property of setting up ammoniacal fermentation. But pathogenic varieties are also found, in the foremost rank of which stand the *cholera bacillus* described by Koch, which was discovered in drinking water in the neighbourhood of Calcutta, and the bacillus of *typhoid fever*; but besides these, others also occur as a contamination of water. Some micro-organisms cannot grow in water alone, as it does not afford sufficient pabulum for their development, but large numbers also perish from being overwhelmed by the growth of the water bacteria.

Very many of the micro-organisms met with in water generate pigment, often in such quantity that considerable volumes appear coloured or fluorescent owing to it, and a few exhibit a brilliant phosphorescence.

Filtration and filters.—Microbes are removed from water by *filtration*, for which purpose use is made of *sand filters* constructed with sand and gravel, *charcoal filters* of plastic carbon, filters of *asbestos*, of *unglazed porcelain*, of earthenware made from burnt *diatomaceous clay*, &c. Förster's filter allows the water to trickle through *sandstone*, and in that of David it is forced in turn through layers of *wool* treated with iron tannate, *sandstone, animal charcoal*, and *gravel*.

The *kaolin* filters on the Chamberland-Pasteur principle consist of porous tubes of porcelain (the so-called ' candles ') about 20 cm. long and 2 cm. thick, closed at one end and provided at the other with enamelled points for the outflow.

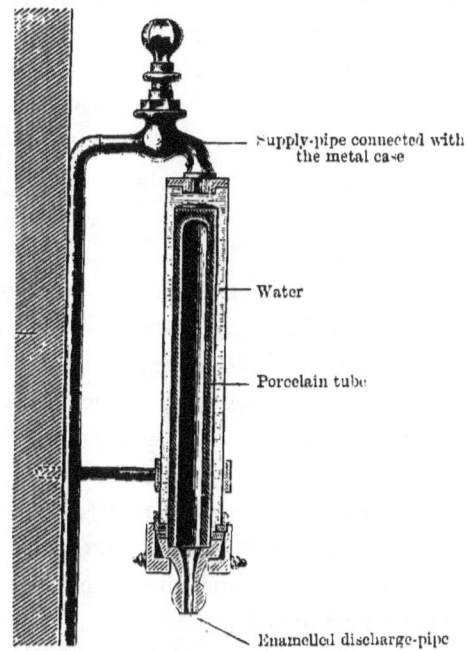

FIG. 44.—KAOLIN FILTER, ON THE CHAMBERLAND-PASTEUR SYSTEM.

These are placed in the water to be filtered or fixed in metal cases and screwed on to the supply-pipes; several may also be connected to form a battery (fig. 44).

The *micro-membrane filter* of Breier consists of a fine netting of metal covered with densely-packed asbestos, which thus forms a thin filtering layer having excessively fine pores.

Variations in water depending on source.—The bacteria which live in water multiply considerably when it has been stagnant for some time, and observations made in flooded districts show that the bacilli of *anthrax, typhoid fever*, and *cholera* are capable of growth on dead portions of plants when moist. Koch has also found the *bacillus of mouse septicæmia* in water, and the *Staphylococcus pyogenes aureus* is not seldom encountered. This is explained by the fact that fresh water contains carbon dioxide; and, moreover, large numbers of germs are often found also in artificial aërated waters, such as seltzer, which contain the gas.

In fresh spring-water the germs are said to sink to the bottom, and hence in some wells which have been disused for a considerable time a larger proportion of germs is demonstrable after the first pumping than later. The micro-organisms are not, as a rule, carried to the well by the ground water, but come from the surface and the superficial layers of soil, and the more ground-water is caused to flow in by constant pumping, the fewer will become the bacteria contained in the water of the spring. When, however, the distance of the spring from the surface is small, or when the well has been made artificially by damming up the earth, or when sewers extend down into the ground-water, then this water in which the spring stands will be very rich in bacteria (Arnold).

A drinking-water which can be termed good from a bacteriological point of view must be poor in fission-fungi, and consequently must not have stagnated in the water-pipes, and there must be security that no communication can take place by crevices and fissures between the reservoir or the mains on the one hand, and drains or sinks on the other. According to Rubner, the natural filtration through the soil under which the spring water lies purifies it so thoroughly that it comes to the light of day in an almost sterile condition, only containing two or three germs per cubic centimeter.

Examination of water.—For purposes of examination ½ to 1 c.cm. is taken with a sterilised pipette and mixed with sterile melted gelatine, which is then poured upon a plate, and the development of the colonies carried on at the temperature of the room. The number of islets formed is then ascertained with the aid of a counting apparatus, and in this way the relative value in micro-organisms of various samples of water is determined.

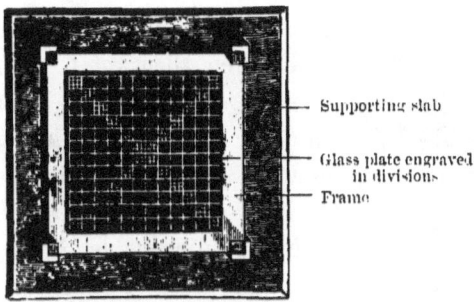

FIG. 45.—WOLFFHÜGEL'S COUNTING-PLATE.

The *counting apparatus* (Wolffhügel's counting-plate, fig. 45) consists of a black slab upon which the plate with the gelatine culture is laid, and over this is arranged a pane of glass on which squares of uniform size have been engraved. The islets in the individual squares are then counted with the help of a lens, and an average struck, when the number so obtained multiplied by the total number of squares on the plate gives approximately the total number of colonies for a certain area, a number which varies with different kinds of water. The water to be used in this experiment must not be kept, but must be examined immediately after collection. In examining water presumably rich in germs—for example, that from rivers or ponds—the volume of water used for observation must be diluted with sterilised distilled water (generally in equal parts or in the proportion of one to nine), as otherwise the colonies lie so close together that they cannot be counted, or else they liquefy the gelatine too speedily.

The counting apparatus can be rendered complete by cutting a piece of some size from the upper part of a square box, and placing a plane mirror obliquely in the interior. If the gelatine plate be now laid upon the box, the number of islets on each square can easily be ascertained by the transmitted light.

Pfuhl's method.—If the examination can be carried out immediately at the spring, the water to be analysed is poured into sterilised vessels, which are at once closed with a sterilised plug of cotton-wool. To obtain the water without catching extraneous germs, Pfuhl uses flat-bottomed glass tubes partially emptied of air, and having the ends drawn out into capillaries, bent at a right angle, and sealed. The points are broken off actually at the spring, and the tubes filled with water and again sealed. For the purpose of transport small cylindrical glass bottles, provided with ground-glass stoppers, are used, which have been sterilised and covered with india-rubber caps. To collect the water from a delivery-pipe the cap and glass stopper are removed, the bottle completely filled, and carefully closed again. The water which first flows away, however, must not be used for examination. To obtain water from a spring the rubber cap is taken off, but the stopper is only removed under the surface of the water, which is allowed to flow into the flask for about a minute, and then the bottle is closed again and lifted out and the rubber cap drawn over the glass stopper.

Kirchner's method.—About 36 cm. length of glass tube, of the diameter commonly used for making connections between apparatus, is bent to a U form in the flame, and both ends are drawn out to points, after which it is sterilised and sealed hermetically at both extremities while still hot. At the place where water is to be collected both ends are broken off, and one held in the water while suction is exerted at the other until the tube is full, when both points are sealed on the spot. The tubes are sent packed in ice.

Other methods.—To gain some preliminary information as to what micro-organisms are present, a few drops of the water to be examined are evaporated on a cover-glass, which is drawn several times through the gas flame to fix the dry residue and covered with one or two drops of a staining solution. In washing off the stain, the stream from the wash-bottle should not be directed on the actual deposit from the water.

It is advisable that every examination by culture should be preceded by sedimentation, for which purpose Finkelnburg has contrived an apparatus consisting of a cylindrical vessel with a bottom capable of being lifted out. Water being allowed to drop in through an opening in the bottom which can be closed by a glass tap, the floating particles gather on the movable bottom of the glass, and in this way a deposition of the organised impurities can rapidly be obtained. Csokor's or Gärtner's centrifugal machine serves this purpose still better. In order to isolate the bacilli of cholera and typhoid and other bacteria endowed with motility, Ali Cohen has brought forward a peculiar method depending on the *chemotactic* action of certain stimulating substances, especially of the juice of raw potatoes. A small glass capillary tube is filled with the fluid found on the cut surface of the raw potato, and is sealed at one end. A ridge of paraffin is now made on a microscope slide, enclosing a space into which the fluid to be examined is introduced, and the sealed end of the capillary tube is fixed in the paraffin ridge while the open end extends into the fluid. The whole is now protected with a cover-glass, when it can be seen under the microscope that none but the motile micro-organisms make their way into the interior of the tube. They can easily be isolated afterwards by cultivation on plates.

Micrococcus aquatilis.—According to Bolton, this microbe is one of the commonest inhabitants of water. The cocci are very minute, and are usually grouped in irregular clumps. Their growth does not liquefy gelatine, on the surface of which there develop circular deposits with a gloss like that of porcelain, from the centre of which furrows radiate out, so as to give the colony the figure of a liver acinus. In thrust-cultures growth takes place both on the surface and along the needle-track. A white coating develops on agar.

Micrococcus agilis, found by Ali Cohen in drinking-water, is met with in the form either of diplococci or streptococci, which possess the power of lively automatic movement. It liquefies gelatine very slowly, so that an evaporation of the fluid along the thrust canal often takes place within three weeks, leaving a dry funnel-shaped cavity. It forms a rose-red deposit both on agar and potato.

Micrococcus fuscus, described by Maschek, consists of immotile cocci which frequently have an elliptical form. Round light- or dark-brown colonies appear on the gelatine plate and speedily liquefy, and in the canal of a thrust-culture liquefaction also progresses with tolerable rapidity, a sepia-brown pellicle forming on the surface of the fluid. The slimy deposit on potatoes is also distinguished by a brown colour.

Micrococcus luteus.—This, which was described by Cohen, consists of small immotile elements, forming a rather flocculent zoogloea. It appears in irregular colonies on the gelatine plate, while in thrust-cultures a yellow deposit is seen on the surface, and granules form along the track. The gelatine is not liquefied. A slimy coating develops on potatoes and agar, and prominences and hollows appear on old cultures. The yellow pigment shows itself capable of resisting the action both of acids and alkalies.

Micrococcus aurantiacus was also discovered by Cohen, and consists of small immotile elements, which sometimes occur in the form of diplococci. The colonies on plates as well as thrust-cultures in gelatine show a fine orange-yellow colour, and the growths on agar and potatoes are also beautifully tinted. Gelatine is not liquefied.

Micrococcus fervidosus.—Adametz has described the *Micrococcus fervidosus*, which consists of small elements whose colonies are first seen on the gelatine plate as dots of a pale yellow colour, becoming brown later. In gelatine thrust-cultures a granular growth appears along the canal and a thin coating on the surface. Superficial cultures on agar show a gloss like that of mother-of-pearl, and a dirty white deposit occurs on potato. Abundant bubbles of gas are disengaged on glycerine jelly, but there is no liquefaction of the gelatine.

Micrococcus carneus.—This micro-organism, described by Zimmermann, is distinguished by the cluster-like arrangement of its elements, which are immotile. Round reddish-coloured colonies appear on the gelatine plate, but in older cultures the red tint fades towards the circumference. In thrust-cultures the colour only appears on the surface. A flesh-red, or sometimes violet layer develops on agar and potato.

Micrococcus concentricus.—This, like the preceding, was found by Zimmermann in the Chemnitz water-supply. The cocci are arranged in clumps, and occur on gelatine plates in the form of blue-grey dots, while thrust-cultures in gelatine show on the surface a greyish-brown

K

disc notched in a radiating manner at the margin, round which a light brownish circle runs, and this again is surrounded by a second circle of a brightly-shining appearance, so that the surface of the gelatine appears marked with concentric rings. A dirty grey deposit forms on potato.

Diplococcus luteus.—This diplococcus, found by Adametz, forms rather long chains, whole pieces of which move about in a lively manner in the hanging drop. It liquefies gelatine slowly and forms round brownish-yellow colonies which spread with tolerable rapidity. In thrust-cultures the growth is more active on the surface, showing lemon-yellow deposits in concentric layers. It appears on agar as a yellow or brownish-red coating, while that on potato is dirty yellow and exhales a mouldy odour. Caseine is precipitated from milk by cultures of the micro-organism.

Bacillus fluorescens liquefaciens and Bacillus nivalis (glacier-bacillus).—These both display short, thick rods possessing the power of easy movement. Fick found the *Bacillus fluorescens liquefaciens* also in the conjunctival secretion, and Schmolk found the *Bacillus nivalis* in glacier ice. Both form on the gelatine plate colonies which have a funnel-shaped hollow in the centre, and exhibit a greenish-yellow fluorescence. Thrust-cultures grow slowly in the deeper part, but somewhat more rapidly in the case of *Bacillus fluorescens liquefaciens* than in that of *Bacillus nivalis*, and the former shows on the surface a depression resembling an air-bubble, owing to evaporation, whereas with the latter the liquefaction extends over the gelatine. The non-liquefied portion of the medium shows a greenish-yellow fluorescence. A whitish layer appears on agar along the inoculated streak, and the mass becomes fluorescent. The deposit on potato is yellowish-brown.

Bacillus fluorescens non-liquefaciens.—A bacillus closely related to the above-described fluorescent but liquefactive bacilli has been discovered in the *Bacillus fluorescens non-liquefaciens*, small rods destitute of motility which form on the gelatine plate shimmering colonies with indented edges, having a darker spot in the centre, and a lighter coloured leaf-like figure all round. In thrust-cultures there is a superficial growth of considerable vigour, but nothing can be observed along the needle-track; the shimmer, however, pervades the whole of the gelatine. This bacterium is distinguished from the *Bacillus erythrosporus* by the fact that the latter shows red-coloured spores.

Bacillus erythrosporus was found by Eidam in drinking-water and in various putrefying albuminous fluids. The rods are slender, have rounded corners, and are actively motile, and the cultures are characterised by the development of a dichromatic pigment, appearing orange-yellow by direct, green by transmitted light. The colonies on the gelatine plate are circular, and show in the centre a darker spot around which spreads a wider light zone. Round every colony fluorescence appears, and soon spreads over the gelatine, so that the entire plate exhibits the phenomenon. Thrust-cultures show a growth upon the surface from which the fluorescence advances deeper until it extends along the whole track of inoculation. A reddish colour, becoming later nut-brown, develops on potato. The spores are characterised by a red gleam, which gives them the appearance of being stained with fuchsine.

Bacillus arborescens has been frequently detected by P. Frankland in the water-supply of London. Its rodlets are thin and motile, and it appears on gelatine plates in iridescent colonies resembling a trunk with its branches, the latter being arranged in sheaves. A superficial iridescence is visible likewise in the thrust-cultures, and slow liquefaction soon sets in; while on agar and potato a yellow or orange-coloured growth develops, with an iridescent margin.

Bacillus violaceus.—In water of a violet tint, in that from rivers and water-works—in the Thames and the Spree, for instance—there are found motile rods with rounded ends, whose colonies on the gelatine plate are at first small, resembling air-bubbles enclosed in the gelatine, but later on liquefy it and form granular islets of a bluish-violet colour, which are distinctly visible as early as the fourth or fifth day. In the thrust-culture a funnel-shaped area is formed by the liquefaction, to the bottom of which the blue-coloured masses of bacilli sink. A dark blue coating develops on agar, and the bacillus grows very well on potato, spreading out radially from the place of inoculation until the disc is nearly covered. Blood serum and plovers' egg albumen are liquefied with the formation of a violet colour.

Bacillus gasoformans.—The gas-forming bacillus consists of small highly motile rods, and forms islets on the gelatine plate, which, though at first small, rapidly liquefy the medium and spread into its substance as well as over the surface. In this way a capsule-like figure is formed, in which bubbles of gas are visible; but the formation of these bubbles is especially characteristic in the clear gelatine along a thrust-canal. The gelatine is speedily liquefied. Growth only takes place at the temperature of an ordinary room.

Bacillus phosphorescens.—Fischer found the **Bacillus phosphorescens indigenus**, or native luminous bacillus, in phosphorescent water from Kiel harbour. Its rods are short, rounded at the ends, and actively motile, and will not grow on serum nor on potato. Small round colonies form on the gelatine plate, which liquefy rapidly, and in about eight days present the appearance of holes cut with a punch in the gelatine, and evidently containing air. The young colonies are sea-green, the older of a dirty greyish-yellow, forming variously shaped flakes. The thrust-culture is of a slightly conical or sand-glass form, with a thin deposit on the edge of the gelatine; but in old cultivations the colonies become heaped on the bottom without fluid. The light given off by them is of a bluish-white colour, and if traces of the culture be added to sea-water, it acquires the property of phosphorescence like that observed at sea.

Closely allied to the foregoing is the **Bacillus phosphorescens indicus**, or Indian luminous bacillus, which was also found in sea-water by Fischer, and which forms small, thick, energetically motile rods. Round, sharply-defined colonies develop on the gelatine plate about the third day, and are also of a sea-green colour with a rosy shimmer, but later turn dirty yellow. The thrust-culture shows on the fourth day a cavity filled with air on the site of the puncture, and later the loss of substance increases, the liquefied gelatine being covered by a dirty yellow pellicle. A white layer develops on potato. Boiled fish and meat furnish a good medium for the growth of the micro-organism, which is checked at temperatures below 10° C. The fact of its growing on potato distinguishes it from the indigenous species. In the presence of air and moisture it shows a luminosity in the dark, best at a temperature of 25° to 30° C. Both bacilli admit of being photographed by their own light. They grow best on *herring gelatine*, which is prepared in a manner similar to ordinary gelatine.

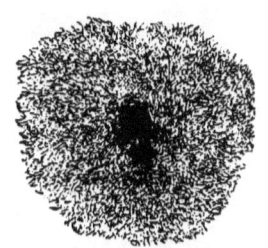

FIG. 46.—ISLET OF BACILLUS RAMOSUS ON A GELATINE PLATE, IN PROCESS OF LIQUEFACTION.

Bacillus ramosus, or root-bacillus (*Wurzelbacillus*), shows short rods with rounded ends and possessing slight power of movement, and is found in the water both of wells and rivers, and also on the surface of the ground. Small colonies with ill-defined margins develop on the gelatine plate, and later on send out shoots resembling the filaments

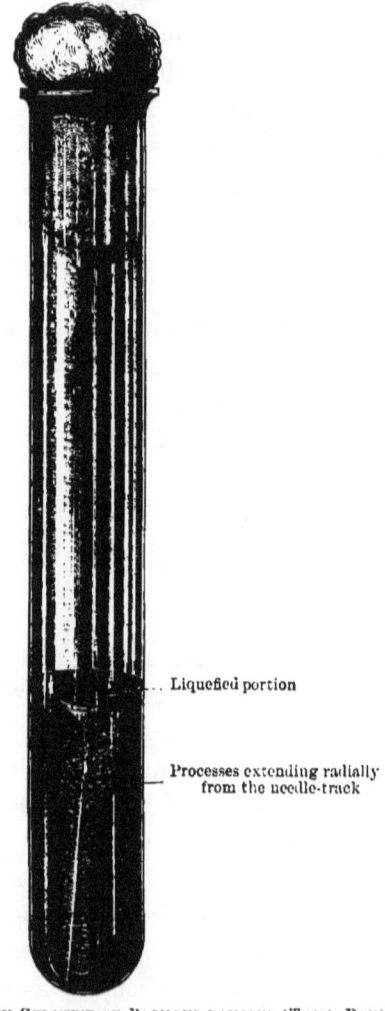

Fig. 47.—Thrust-Culture in Gelatine of Bacillus ramosus (Third Day).

of mould. When, after growing for some time, the fibres have become entangled with one another, they form an interlacing root-like network (fig. 46), an appearance which is seen still more characteristically in the thrust-culture, processes springing out radially from the needle-track, after the manner of fibres from a tap-root. The aspect of the culture has been aptly compared to that of an inverted pine-tree (fig. 47). Older cultures are completely liquefied, and carry a pellicle on the surface. The ramification is equally distinct upon agar, and spores are also formed on potato. On plovers' egg albumen there soon develop whitey-grey colonies growing in root form, which become continually more and more interlaced and entangled. The liquefaction of the albumen is very tardy.

Bacillus aurantiacus.—This bacillus, found by P. Frankland in deep wells, shows only a feeble motility. The rods are often arranged in pairs, and grow out into threads. Gelatine is not liquefied. Small raised colonies appear on the plate, which are at first dark-coloured but subsequently become bright orange-yellow, and this orange colour is also very distinct on the surface of thrust-cultures, whereas the track of the thrust shows no growth. On agar the deposit is also of an orange colour. On potato the growth is restricted to the spot inoculated.

Bacillus aureus.—Adametz found the *Bacillus aureus* in water, and Unna upon the skin of persons suffering from *seborrhœic eczema*. The rodlets are slender, have but slight motility, and do not liquefy gelatine. The colonies on the gelatine plate are irregular, coalesce with each other, and develop a golden-yellow pigment, which also occurs in the form of hemispheres on the surface of the streak-culture. On potato the golden-yellow colour soon changes to a brownish red.

Bacillus bruneus.—The brown bacillus described by Adametz and Wichmann shows small irregular rods. The colonies, which do not liquefy the gelatine in their growth, are rounded and whitish at first, later brown. The brown colour is also developed along the whole length of the thrust-canal in the gelatine.

Bacillus aquatilis, which was described by P. Frankland, exhibits short, feebly motile rods. On gelatine plates the colonies develop in the deeper parts, and spread thence to the surface, liquefying the gelatine. From the surface bundles of filaments extend to the periphery. In thrust-cultures a weak yellowish coloration first shows on the surface, and liquefaction takes place comparatively late. The growth on agar is restricted to the line of inoculation, and on potato the colonies show a similarly meagre development. The bacillus was found in the water of deep wells sunk in a chalky soil.

Bacillus aquatilis sulcatus.—Weichselbaum found four micro-organisms in the water of the high-level water-works of Vienna at a time when water from the Schwarza was laid on,[1] and these he designates as *Bacillus aquatilis sulcatus* 1-4. The first three forms have motile elements, those of the fourth variety are shorter, and do not move. On the gelatine plates colonies develop which are thicker in the centre than at the periphery, and possess a distinctly coloured border, while on their surface furrows can be seen with a low power crossing one another at the most varying angles, and this network of wrinkles is denser in the centre than at the periphery. The points of distinction of these varieties are limited to variations of temperature and differences of odour, which is like that of whey in the first form, resembles urine in the second, and is exceedingly unpleasant in the third, while the cultures of the fourth variety give off no smell. A superficial growth appears

---

[1] The drinking-water supplied to Vienna is brought from the Schneeberg, a distance of 100 kilometers. In summer, however, sufficient cannot be obtained from this source, so that other water must be laid on in addition, and this is derived from the Schwarza, a little river in the vicinity of the Sœmmering.

in thrust-cultures. The fourth form does not grow on potato.

Bacillus aquatilis radiatus.—Zimmermann found this bacillus in the water of the Chemnitz water-works. The rods are motile and form irregular colonies on the gelatine plate, with root-like processes arranged in a radiating manner. The gelatine is liquefied with tolerable rapidity, and the mass of bacteria is collected in the centre of the fluid colony, while all round it extends a yellowish-coloured ring which is environed by a cloudy yellowish mass. Thrust-cultures show a circular pellicle on the surface when liquefaction has taken place. A layer forms on agar which appears bluish by direct, yellowish by transmitted light.

Bacterium Zürnianum, described by Adametz, displays short, immotile rods. The colonies, which do not liquefy gelatine, form masses like clusters of grapes, and the surface of thrust-cultures has also a similar appearance. It grows on potatoes at 25° to 30° C.

Bacillus membranaceus amethystinus.—Eisenberg and Jolles have discovered this bacillus in the spring-water of Spalato. It has the power of liquefying gelatine with development of a dark violet pigment, and its rods are short and immotile. Small, dark violet colonies develop on gelatine plates and liquefy slowly, while a violet pellicle appears on the surface of the liquefied mass, resembling a membrane stained with gentian violet. A deposit with serrated edges develops on thrust-cultures and acquires a violet colour in about a fortnight, slowly liquefying the gelatine. The liquid mass is covered with a violet pellicle in this case also. A whitish layer, which later becomes violet, forms upon agar; the deposit on potato is of a dirty olive-green, while a violet sediment and a violet pellicle develop in bouillon.

Bacillus indigoferus.—The bacillus of this name, detected by Claessen in the water of the Spree, has slender rods with rounded ends and showing vigorous motility, but which often adhere together and form entire clumps, and seem to possess a jelly-like envelope. The colonies, which do not develop on the gelatine plate until the third day, contain an indigo-blue pigment as early as the day following. In thrust-cultures a punctiform mass of a deep indigo blue appears round the site of inoculation even in twenty-four hours, and the gelatine is not liquefied. On agar an indigo-blue layer develops over the extent of the line of inoculation, its surface being covered with a shining pellicle. On potato an intensely deep indigo blue develops in three or four days, but only when the reaction is acid. The formation of pigment does not seem to depend on the presence of light.

Bacillus ianthinus was found by Zopf in the Chemnitz water-supply. It consists of short rods with active motile power, which liquefy gelatine, and form a bluish-violet pigment.

Bacillus ochraceus.—Zimmermann obtained from the same source as the foregoing a bacillus with ochre-yellow pigment, which has been described as the *Bacillus ochraceus*. Gelatine is slowly liquefied. The rods show little motility.

Bacillus gracilis.—Zimmermann described a micro-organism to which this name has been given on account of the form of the rods, which are slender and fine and possess an oscillating rotatory motion. Gelatine is liquefied. On agar a bluish-white pigment appears. There is but little growth on potato. It has the property of thriving better in the substance of the solid media than on the surface; hence it will also develop under the mica plate (see p. 61).

Bacillus sulphydrogenus.—Miquel found not only in drinking-water, but in rain and canal water, an anaerobic bacillus possessing the peculiar property of decomposing albumen with the formation of sulphuretted hydrogen. Its cells are very short motile rods, which increase in length when grown artificially. Carbon dioxide and hydrogen are liberated from a nutrient medium which contains no sulphur. 1 grm. of sulphur is decomposed in forty-eight hours in a bouillon mixed with ammonium tartrate and with sulphur in excess.

Other bacteria, however, also generate sulphuretted hydrogen, to prove the presence of which it is only necessary to hang strips of paper saturated with lead acetate in the test-tubes containing the cultures, when a blackening of the paper will be found with most water and air bacilli. According to Macé this reaction becomes more distinct if some flowers of sulphur be added to the culture medium.

Bacillus of Asiatic cholera, and allied micro-organisms.— Koch discovered a comma-bacillus in the evacuations and the intestines of cholera patients, and adduced evidence proving that this micro-organism is the cause of the disease.

FIG. 48.—CHOLERA BACILLI, FROM A PURE CULTURE. (After Jaksch.)

He found it also in the water of a tank in the neighbourhood of Calcutta. Experiments have shown that the comma-bacilli can multiply very well in sterilised water, whereas in ordinary water they are soon overgrown by the water-bacteria; and they can tolerate just as little the concurrence of other bacteria, and consequently perish in decomposing liquids, although Gruber has obtained pure cultures from alvine evacuations which had been allowed to putrefy for a week.

The pathogenesis of this bacillus has, however, recently been rendered doubtful by the experiments (performed on themselves) of Pettenkofer and Emmerich at Munich, and Hasterlik in Vienna; and of Dr. A. J. Wall, of H.M.

Indian Army, in London, who swallowed cultures of the bacilli.

Cholera bacilli appear as plump rods curved in the direction of their long axis so as to resemble a comma in figure, hence the name 'comma-bacillus' (fig. 48). They have a twist in addition to this curve, so that they represent a kind of spiral bacteria, and on this account have been described as *Vibrio* or *Spirillum cholerae*. The bacteria lie connected in chains forming a tolerably steep spiral, and when examined in the hanging drop show an exceedingly lively motion, so that they might not inaptly be likened to a swarm of midges dancing all over the field. Löffler found a flagellum at one end of the cell. Koch was unable to detect spores in them, but Flügge, on the other hand, has succeeded in demonstrating the formation of arthrospores. They offer but small resistance to the influence of chemical substances, and are destroyed by the acid of the gastric juices; and they refuse to grow upon feebly acid gelatine. They also perish at temperatures above 50° C., but grow at room temperature as well as in the incubator. Drying causes speedy loss of the power of development. They stain in about ten minutes in diluted alcoholic solutions of the aniline dyes, but the process is accelerated by heating the solution, which also increases the intensity of the stain. Special importance is to be attached to the fact that they are decolorised if Gram's process be employed.

Staining is done as follows: An alcoholic solution of fuchsine or methyl violet is diluted with water sufficiently to form a fairly strong aqueous staining fluid, which is poured into a watch-glass. A trace of the pure culture or a minute flake from the intestinal contents is next placed upon a cover-glass and spread out by rubbing lightly with another laid over it, after which the two glasses are drawn apart, dried in air, grasped with a forceps and passed several times through a

STAINING OF CHOLERA BACILLI 141

flame to fix the micro-organisms, and then laid on the surface of the staining solution with the infected side downwards. The solution is allowed to act for ten minutes on the preparations, with, or even without, slight warming, and it need hardly be said that care must be taken to exclude all substances liable to decompose the stain. The cover-glasses are next taken from the solution with the forceps, washed in water, and dried with the prepared surface uppermost, after which a drop of Canada balsam is laid upon the prepared face of the glass, which is in this way cemented to the slide. Microscopic examination is carried out with the help of the oil-immersion lens and Abbé's condenser.

When a sample of the material to be examined for cholera bacilli is diffused through gelatine, and a plate-culture made, colonies are soon obtained possessing in-

Non-liquefied marginal portion

FIG. 49.—ISLET OF BACILLUS CHOLERÆ ASIATICÆ ON A GELATINE PLATE, IN PROCESS OF LIQUEFACTION.

dented and bowed margins, and presenting, when magnified 100 diameters, an appearance as if strewn with bits of glass. After a time, small round excavations appear, corresponding to the colonies, and soon give the entire gelatine the aspect of pock-marked skin, the funnel-shaped cavities being due to evaporation of water from the slowly-liquefying gelatine. The liquefaction does not extend very far at first (fig. 49).

In thrust-cultures also the gelatine liquefies very slowly, the liquefaction being chiefly seen on the surface, and in this case too evaporation of the fluid occurs, so that a bubble communicating with the external air appears in the upper part of the funnel-shaped excavation. From this bubble a thin prolongation runs down along the track of

142   BACTERIOLOGY

inoculation, and comes to resemble a capillary tube as the liquefaction advances downwards and the evaporation pro-

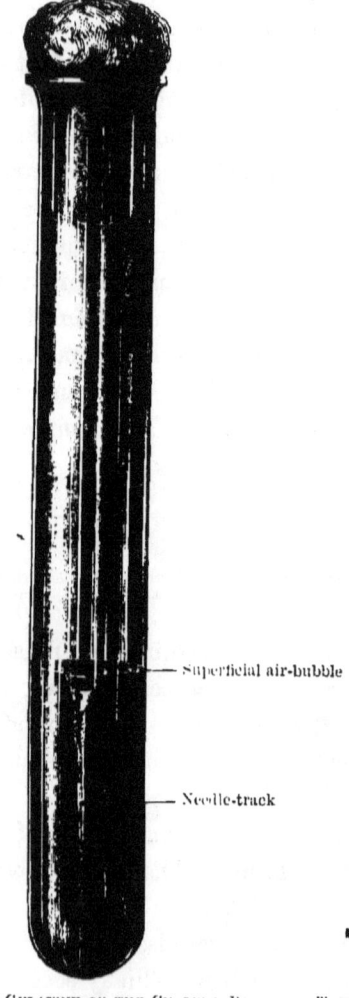

Fig. 50.—Thrust-Culture in Gelatine of the Cholera Bacillus (Third Day).

gresses. When liquefaction has gone still further the bacilli sink in the needle-track, and the coherent mass of bacteria forms a loose thread which is sometimes thicker and sometimes thinner, and which extends to the bottom of the funnel-shaped liquid cavity (fig. 50). In four weeks liquefaction has made such progress that the entire mass of gelatine is fluid.

Cholera bacilli grow with a fair degree of luxuriance upon other media also, forming on culture bouillon a wrinkled, much-folded membrane, while the bouillon itself remains tolerably clear. Only on shaking do the masses rise from the bottom and render it cloudy.

They grow also in sterilised milk, but in milk which has not been freed from germs they undergo speedy destruction, owing to the occurrence of acid fermentation.

In sterilised water they grow rather actively and maintain themselves for a considerable time, so that cholera bacilli find in many places the conditions required for their increase.

On agar the bacilli grow in the form of a whitish layer spreading out from the line of inoculation.

On potato they thrive even when the surface shows a slightly acid reaction, but only at 30° to 40° C. In the region surrounding the site of inoculation there develops a greyish-brown layer, which gradually spreads and presents an appearance almost identical with that exhibited by cultivations of the *glanders bacillus*.

Blood serum is slowly liquefied by the growth of cholera bacilli.

The inoculated streak in the case of plovers' egg albumen (Von Hovorka and F. Winkler) becomes covered with a coating which reflects more light than the surrounding mass of transparent white of egg, and consequently looks lighter. If examined with a lens the streak, which slowly widens, exhibits closely-packed, distorted colonies with a grey sheen, partly united with one another. There is no liquefaction of the nutrient medium.

During the progress of the disease the bacilli are found principally in the mucous membrane and contents of the intestinal canal, but not in the blood.

To examine cholera stools for bacilli, the evacuation is mixed with an equal bulk of alkaline meat-bouillon and let stand in an open glass for twelve hours at a temperature of 30° to 40° C. An abundant development on the surface results, and preparations are obtained by this method of investigation which consist of cholera bacilli only. To recognise the presence of the bacilli even without the microscope, a good way is to add 10 per cent. hydrochloric acid, which after only a few minutes stains the cholera cultures a rose-violet colour, forming the *cholera reaction*.

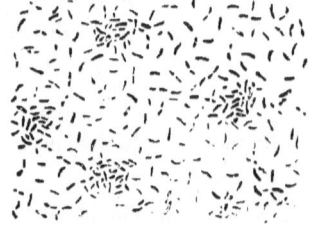

FIG. 51.—FINKLER-PRIOR BACILLI FROM A PURE CULTURE. (After Jaksch.)

The introduction of pure cultures in bouillon into the stomach of guinea-pigs resulted in Koch's hands in an unmistakable infection, when combined with the administration of sodium bicarbonate and opium. To avoid the detrimental action of the gastric juice, and at the same time to diminish intestinal peristalsis, Nicati and Rietsch ligatured the ductus choledochus and made the injection directly into the duodenum. Introduction of cholera cultures into the circulation also causes the death of the animals, and in that case the bacilli can be detected in all the organs.

The problem frequently arises, to determine with certainty whether in a given case it is the bacilli of cholera Asiatica that have to be dealt with, or other micro-organisms resembling them in form. One of these is the *Vibrio proteus*, which was discovered by Finkler and Prior in the evacuations of persons suffering from *cholera nostras*, and which may possibly be identical with the comma-bacillus found by W. D. Miller in carious teeth.

The Finkler-Prior bacillus, or Vibrio proteus, is somewhat larger and thicker than the bacillus of Koch, but the spirilla formed by it are never so long as the cholera spirilla (fig. 51). The culture on a gelatine plate liquefies so rapidly and extensively that the difference between it and a culture of cholera bacillus must at once become apparent. If two gelatine tubes are inoculated by making in them two thrusts parallel with one another, the one with the bacillus of cholera Asiatica, and the other in the same way with *Vibrio proteus*, speedy liquefaction will be observed in the latter, whereas with cholera Asiatica it occurs very slowly and shows at the surface an excavation occupied by a bubble of air. Furthermore the liquefaction of *Vibrio proteus* extends widely into the surrounding parts, the masses of bacilli sink to the bottom, and there is obtained in the liquefied area corresponding to both thrusts the form of an empty stocking-leg (the so-called '*trouser-leg culture*,' fig. 52). Superficially the liquefaction of the gelatine soon causes the two thrust-cultures to coalesce, and a skin then forms upon the surface.

The *Vibrio proteus* grows on potato even at ordinary temperatures, whereas the *cholera bacillus* only grows at that of the incubator.

On plovers' egg albumen a very distinct difference quickly becomes apparent, liquefaction beginning in the case of *Vibrio proteus* as early as the second day, and an intense yellow coloration of the entire nutrient mass soon setting in, whereas the culture of cholera Asiatica shows neither the one nor the other. The nutrient mass runs round from the sides of the test-tube towards the bottom, where it gathers in a comparatively thick stratum (Von Hovorka and F. Winkler).

Deneke's comma-bacillus is more difficult to distinguish from that of cholera. It was grown by Deneke from old

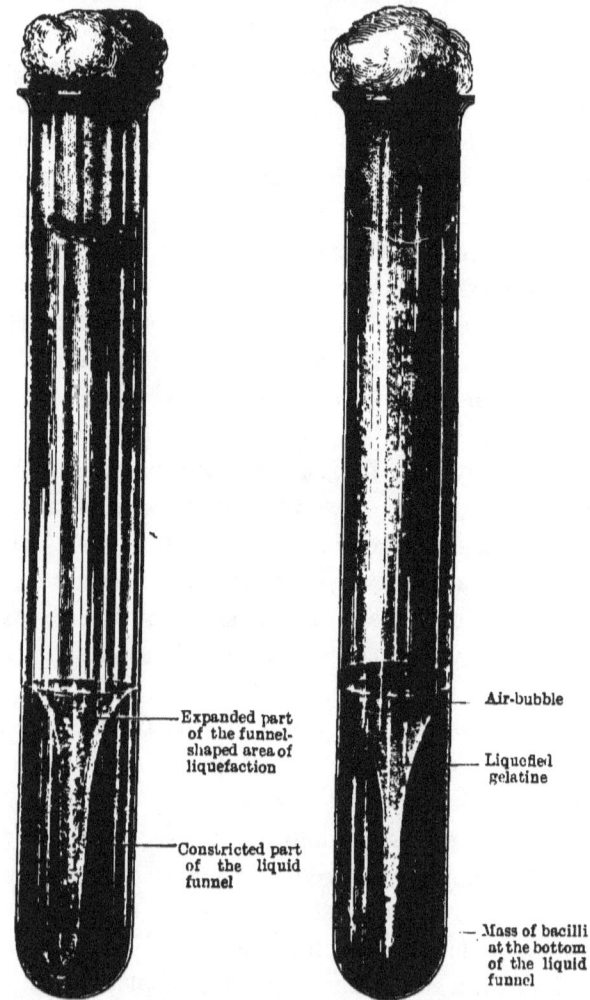

FIG. 52.—THRUST-CULTURE IN GELATINE OF THE FINKLER-PRIOR BACILLUS (THIRD DAY).

FIG. 53.—THRUST-CULTURE IN GELATINE OF DENEKE'S COMMA-BACILLUS. BEGINNING OF THE THIRD DAY. (After Baumgarten.)

cheese, and scarcely differs from the cholera bacillus in its morphological relations. It is, however, distinguished by its speedier liquefaction on the gelatine plate, and the yellowish colour of the colonies, which appear irregular under the microscope, and are surrounded by a thick rampart. Like the cholera bacillus, it shows an air-bubble in thrust-cultures at the uppermost part of the funnel-shaped excavation, but the bubble is larger than in the former case (fig. 53). A delicate yellowish coat, composed of beautifully-formed spirilla, appears on potatoes at incubation temperature. Like the *Vibrio proteus*, this bacterium possesses pathogenic properties.

The differential diagnosis of the *Bacillus neapolitanus* (p. 120), regarded by Emmerich as the originator of cholera, presents no difficulties, as its elements are short immotile rods and the gelatine is not liquefied by their growth, which takes place more chiefly upon its surface.

The **Vibrio Metschnikoffi**, which was found by Gamaleia in the intestinal contents in a Russian disease of poultry, is a curved bacterium forming screw-shaped spirilla of considerable length, but which is shorter and thicker than the cholera bacillus. Its accurate identification is often difficult, as, although the gelatine of plate-cultures becomes fluid at times as rapidly as with *Vibrio proteus*, the liquefaction is sometimes greatly protracted. The thrust-culture shows a distinct air-bubble, which gradually enlarges, and disappears when liquefaction is complete. It grows on potato as a brown deposit, but only at the temperature of the incubator.

A peculiarly characteristic property of the cholera bacillus lies in the *cholera reaction* already alluded to, which was discovered by Bujwid and Dunham. Cultures of the bacilli in media containing peptone (bouillon or gelatine) give a reddish-violet or purple-red colour in a short time when treated with pure hydrochloric or sulphuric acid, in which process a definite pigment, *cholera red*, is developed. Salkowski explains this as an *indol reaction*, since that substance gives a red coloration with nitrous acid, the theory being that the cholera bacilli split off indol from the peptone of the nutrient medium, and at the same time develop nitrites which are decomposed by the addition of a strong acid. Of all the other morphologically and biologically similar micro-organisms, *Vibrio Metschnikoffi* alone gives this cholera reaction.

According to Gruber and Schottelius, a still more trustworthy point of difference is the fact that the cholera bacilli grow with particular luxuriance in a very dilute bouillon, and develop a wrinkled membrane on the surface in a few hours,

FIG. 54.—TYPHOID BACILLI FROM A PURE CULTURE. (After Jaksch.)

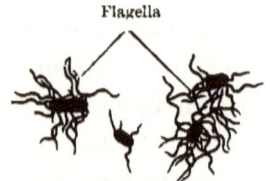

FIG. 55.—TYPHOID BACILLI ('SPIDER-CELLS'). Magnified 1,100 times. (After Löffler.)

whereas the remaining micro-organisms show no such rapid growth. But the most certain diagnostic of all assuredly lies in the manner of growth on plover's egg albumen characterised above.

Brieger was able to obtain some alkaloids and toxalbumins from cholera cultures, amongst others *cadaverine*, *putrescine*, and *choline*. Pouchet also extracted toxines from the actual cholera stools.[1]

Bacillus of typhoid fever.—The *bacilli of typhoid or enteric fever* (the *Typhus abdominalis* of the Germans) are met

[1] [For an account of recent researches on cholera vaccination, see Appendix]—TR.

BACILLUS OF TYPHOID FEVER

with in water as well as in the fæces and organs of patients suffering from the disease. They were described by Eberth and Gaffky, and have been more thoroughly studied by Klebs and Eppinger. The bacilli are short plump rods with rounded ends, the length of which is three times as great as their breadth, and which sometimes unite to form what are apparently threads of considerable length (fig. 54). According to Gaffky and Birch-Hirschfeld they develop spores. The rodlets are distinguished by a high degree of motility, dependent, as Löffler has found, on the possession of flagella, which are present in such abundance that the bacilli, when subjected to the proper staining processes, take on the appearance of spiders (fig. 55).

They thrive whether oxygen is excluded or has free access, although in the latter case the growth is more vigorous, and they stain in watery solutions of the aniline dyes, yielding up their colour, however, very easily on application of different bleaching fluids, so that the demonstration of them in tissues is beset with difficulties; and under this head it is to be observed that they lose their colour completely when treated by Gram's method.

On the gelatine plate there form whitish colonies lying superficially, and at first mere dots, which have an unevenly indented margin. Sometimes the growths lie deeper, and take the form of a whetstone. They soon become yellowish-brown, particularly in the centre of the islets. The gelatine is not liquefied. Thrust-cultures show on the surface a thin growth, which also takes place along the entire inoculated track (fig. 56). A white layer covering the whole surface develops on agar, blood-serum, and plover's egg albumen.

According to Petruschky, the typhoid bacillus is one of those which form acids.

When it is desirable to come to a positive decision

regarding the presence of the typhoid bacillus, the experiment of growing it on potato is indispensable. In three or

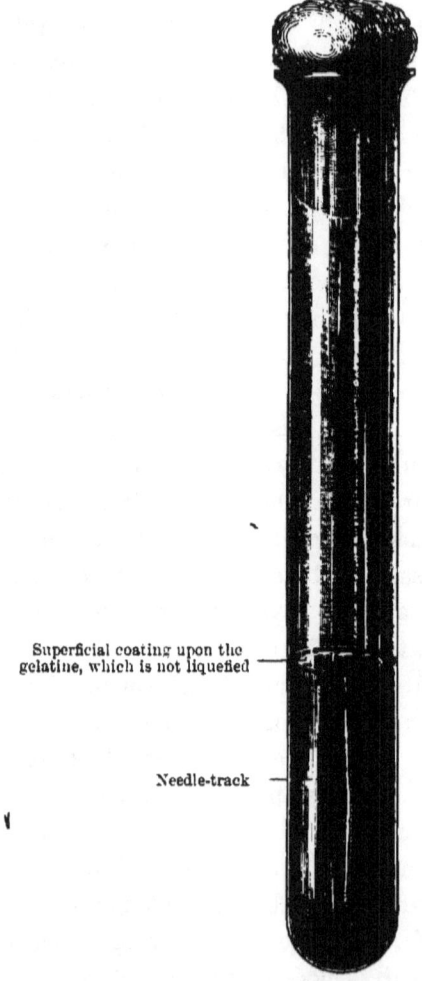

Fig. 56.—Thrust-Culture in Gelatine of the Bacillus of Typhoid Fever. Ninth Day. (After Baumgarten.)

Labels: Superficial coating upon the gelatine, which is not liquefied; Needle-track.

four days at room temperature, and in as little as two at that of the incubator, there appears on the surface a moist, even gloss, although no deposit can be seen even in the part immediately surrounding the site of inoculation. Particles of the surface of a potato showing this appearance exhibit under the microscope micro-organisms shooting hither and thither with the most extreme velocity. These phenomena of growth are of particular importance in order to avoid mistaking them for other species of bacteria which resemble them in all remaining particulars, such as the *bacillus of Emmerich*, which forms a greasy yellowish layer upon the discs of potato. E. Fränkel and Ali Cohen have given prominence to the fact that this growth only occurs upon slices of potato having an acid reaction, so that the reaction of the potatoes must always be previously tested. Fränkel himself, however, as well as others, most recently Kamen, have drawn attention to an atypical growth of the typhoid bacilli upon potato, in which a yellowish layer, which later becomes brown, develops slowly out from the area of inoculation and spreads in a tongue-shaped figure, the potato assuming a violet tint after some days.

As a further means of recognition, Chantemesse and Widal have mentioned a peculiarity of typhoid bacilli, namely, that they thrive on a nutrient gelatine which has been mixed with 2 per 1,000 of carbolic acid, whereas all other micro-organisms perish on this mass.

Rodet has proposed to heat the gelatine, after inoculation with the water to be examined, for from half an hour to two hours in the water-bath at 45° C., by which means at least the troublesome liquefactive germs should be eliminated.

Vincent recommended that a sample of the water under investigation should be transferred to bouillon, kept at 42° C., with which five drops of a five per cent. solution of carbolic acid have been mixed.

Holz prepared an acid gelatine by the addition of the juice of raw potatoes to ordinary nutrient gelatine. Only a small number of indifferent varieties develop on this medium, while the typhoid bacilli grow characteristically.

Parietti adds to several test-tubes, each containing 10 c.cm. of neutral bouillon, from three to nine drops of a hydrochloric acid phenol solution (4 grms. hydrochloric and 5 grms. carbolic acid to 100 grms. water), deposits them for twenty-four hours in the incubator, and then treats them with one to ten drops of the water under examination. If turbidity is visible after another twenty-four hours' standing in the incubator, it may be concluded with certainty that typhoid bacilli are present.

Intravenous injections kill rabbits in about twenty-four hours, when the bacilli may be detected in the urine, blood, and excreta.

Infection takes place principally by means of water contaminated with the bacilli, but also by contaminated milk and linen. The microbes are capable of effectually resisting the action of the gastric juice, and as soon as they reach the intestinal canal they penetrate into the lymphatic canals and are carried by the stream of lymph into the other organs, particularly the [mesenteric glands] spleen and liver. Eberth has shown that they may penetrate into the placenta, and in this way reach the fœtus.

[The typhoid bacilli have also been found in the blood, not only in the rose spots (Neuhauss) but in the general circulation, having been detected in blood from the finger.]

Bacterium coli commune.—Rodet and Roux found this bacterium in the water of localities where typhoid was prevalent, and it has been constantly met with by Escherich in the intestinal canal of suckling infants. It consists of short, slender rods possessing a sluggish motility, which occur sometimes singly and sometimes in pairs, and which are decolorised if treated by Gram's method. Gelatine is not liquefied, the colonies having a tendency to spread out over the medium in a thin superficial film. They show a dull white colour and an irregularly indented border. In thrust-cultures white buttons develop along the needle-track and a delicate film on the surface. The colonies on potato are yellow and juicy, and a white layer appears on serum. The micro-organism is also capable of growth in the absence of oxygen upon nutrient media containing grape-sugar, and then generates a gas consisting of hydrogen and carbon dioxide. Rabbits succumb to a subcutaneous injection in from one to three days, with the symptoms of diarrhœa and collapse.

According to Gasser, an agar medium tinted with fuchsine is decolorised only by this bacterium and the bacillus of typhoid fever, whilst a sufficient point of distinction between these two is, that the growth of *Bacterium coli*

FIG. 57.—SPIRILLUM UNDULA, WITH FLAGELLA. Magnified 800 times. (After Löffler.)

commune remains restricted to the strip inoculated, whereas that of the *typhoid bacillus* forms a tolerably broad streak with very bowed and irregular edges.

Spirilla in water.—Spirilla are found in copious numbers in stagnant water, and are marked by an exceedingly active motility, darting across the field with manifold twists and turns. They often lie together in clumps, which look to the naked eye like flakes of mucus. The individual spirilla possess from one and a half to four turns, or sometimes as many as six, and have some flagella on their ends. They are described as **Spirillum undula** (fig. 57).

Other micro-organisms in water.—A considerable number of the micro-organisms met with in the air find a congenial pabulum in water also, and hence are constantly met with in the various examinations of that medium. Amongst these are found the *Micrococcus radiatus, Micrococcus cinabareus, Micrococcus flavus liquefaciens, Micrococcus desidens, Micrococcus flavus tardigradus, Micrococcus candicans, Micrococcus viticulosus, Staphylococcus pyogenes aureus, Staphylococcus pyogenes albus, Staphylococcus cereus albus, Sarcina alba, Sarcina candida, Bacillus subtilis, Bacillus multipediculosus,* &c., as well as some which decompose milk and will be described under that head.

CHAPTER VII

BACTERIOLOGICAL ANALYSIS OF EARTH AND OF PUTREFYING SUBSTANCES

Micro-organisms in the soil.—The examination of soil proves that very many micro-organisms which thrive in the air and in water can also grow in earth. Moreover, in every putrefactive process on the surface of the ground there occurs an oxidation, or resolution of organic matter with the aid of atmospheric oxygen; consequently, in all these processes a considerable number of micro-organisms are afforded the possibility of maintaining themselves and multiplying. In agriculture the land is manured with the view of enabling highly complex organic substances to undergo decomposition on the surface of the ground into simple combinations, capable of serving as nutriment for plants and of being assimilated by them in order that they may be once more converted into higher combinations. In these putrefactive processes the action of bacteria takes a very prominent part. That it is the *surface* of the soil which is so very rich in varieties of germs becomes evident from the fact that at so short a distance as one to two meters beneath the surface the amount of bacteria present rather suddenly decreases, and that further down, at a depth of three or four meters, the earth is found to be completely free from germs. The ground-water level of the soil is tolerably pure in this respect, and hence also only very few micro-organisms are found in the water of springs. For the same reason fountains fed by pipes, if properly constructed and kept clean, deliver water which is poor in germs, or entirely free from them (C. Fränkel).

According to Kirchner, the freedom of ground-water from germs is due to the filtering action of the soil, and therefore, where this is too coarse and porous, the filtering power fails, and the whole or a part of the germs met with in the upper stratum of earth pass unhindered into the ground-water. According to Reimers, the germs contained in the deeper part grow more slowly than those derived from superficial layers.

Method of examination.—If it is desired to examine soil, or the dust of windows and rooms, for micro-organisms, a small sample is taken—freshly, if possible—and introduced into sterilised nutrient gelatine, which has been melted but is not too hot, in order to prepare a roll-culture by Von Esmarch's method. Cultures of this form are preferable to plates, because the small particles of earth do not sink to the bottom of the tube and get missed, as is liable to happen in pouring out the contents on the plate. Special instruments are employed for obtaining earth from different depths, of which a borer constructed by C. Fränkel is that principally in use.

When the individual islets have been isolated by means of the roll-culture, they are transferred to different plates, in order to obtain pure cultivations of the particular organisms. The examination of anaerobic micro-organisms is carried on by the methods detailed above.

The *brown mould* is very widespread in the earth as well as in air (p. 104), and some micrococci are found which liquefy gelatine. Generally speaking, the cocci are more numerous than the bacilli.

Bacillus mycoides (earth bacillus).—Flügge found, as a very frequent guest in the soil of fields and gardens, a micro-organism whose rods strongly resemble those of anthrax, but are distinguished from them by a lively motility. Gelatine is liquefied. On plate-cultures there

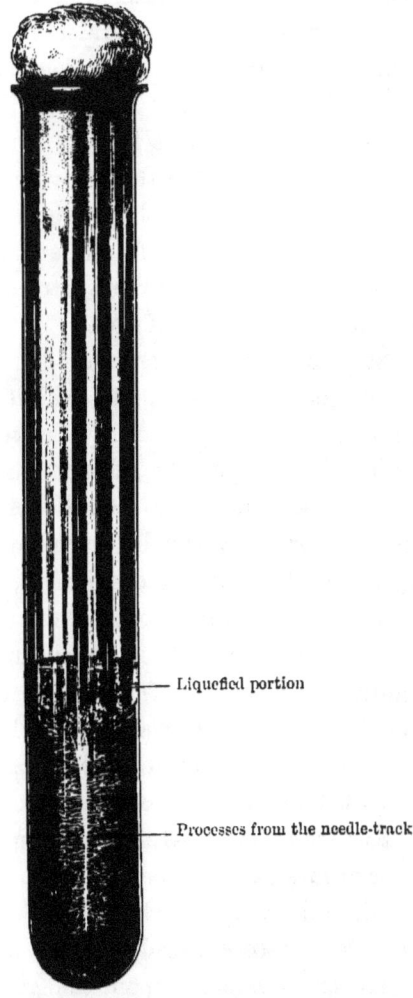

FIG. 58.—THRUST-CULTURE IN GELATINE OF BACILLUS MYCOIDES (FOURTH DAY).

appear colonies which ramify like mycelium, so that the plate looks as if overgrown with moulds. In thrust-cultures liquefaction sets in on the surface as early as the second day, while very delicate fine branching filaments run out from the track of inoculation on all sides, and have a tolerably even length, thus differing from *Bacillus ramosus*, the processes of which diminish in size from above downwards (fig. 58). Ramifications resembling mycelium develop in like manner upon agar, and have at first some likeness to the barb of a feather; but the surface gradually becomes covered with a thick coating. On serum an irregularly-outlined granular layer appears even in twenty-four hours, and spreads in a fern-like manner over the surface. Potatoes are seen after two days to be invested in a fine close mycelium-like texture of fibres.

Bacterium mycoides roseum.—This microbe, described by Scholl and Holschewnikoff, exhibits tolerably large rods which are destitute of motility. A red colour develops early in the colonies upon the gelatine plate, which coalesce with one another and very soon liquefy. Thrust-cultures in like manner show rapid liquefaction, with a red-coloured superficial skin and a red sediment, but without any staining of the gelatine itself. Development takes place at room temperature. Surface-cultures on agar display a beautiful rose colour if grown in the dark, whereas the growth is white if cultivated in daylight. Solutions of the red pigment show an absorption band in the green when placed before the slit of a spectroscope.

Bacillus radiatus.—Lüderitz found the *Bacillus radiatus* in the ground, and in the juice from the subcutaneous tissue of white mice which had been inoculated with garden mould. Its rodlets possess a ready motility and grow anaerobically, liquefying gelatine. Upon the gelatine plate, in 'high' thrust-cultures, and on agar, there appears a tangle of anastomosing fibres, recalling the radiating forms of moulds. A very unpleasant-smelling gas is generated in cultures on serum or sugared gelatine.

Bacillus spinosus was also found by Lüderitz in the juice from the tissues of white mice inoculated with garden mould. Like the last, it can only develop anaerobically, and liquefies gelatine. In high cultures there are visible in two days little punctiform fluid spots, from which radiating processes soon push out; in later stages an expansion appears at the layers of gelatine above and below the track of inoculation, giving the slimy, liquefied mass the form of a sandglass. This bacillus also generates a gas, which has an odour resembling cheese in growths on gelatine containing sugar.

Bacillus liquefaciens magnus.—In the juice of the subcutaneous tissue of animals inoculated with garden mould, Lüderitz also found the bacillus of this name, which consists of large rods rounded at the ends, and endowed with an active motility. It is also an anaerobe, and its growth resembles that of the moulds and liquefies gelatine, the liquefaction in thrust-cultures not taking an hour-glass-, but a more cylindrical sausage-like, form. Mossy colonies develop on agar. The gas generated by it smells very unpleasantly, recalling the odour of onions.

Bacillus scissus.—This bacillus was discovered in the earth by Percy Frankland. It displays short, thick, immotile rods, and does not liquefy gelatine. The colonies on the gelatine plate develop superficially, and in thrust-cultures no growth takes place along the puncture canal, but an irregular deposit with smooth edges forms on the surface. A slight greenish colour is imparted to the gelatine.

Besides the above there exists in the earth an entire series of different micro-organisms, amongst which especially the *Bacillus ramosus* and *Bacillus subtilis* are met with. They possess, however, no significance, so far as research has shown up to the present.

Clostridium fœtidum.—By the name *Clostridia* are understood those forms of bacillus which develop spores in the centre, so that, owing to the bulging there and tapering of the ends, figures of a distinctly spindle shape are formed (see fig. 1).

The *Clostridium fœtidum* is, according to Liborius, an uncompromising anaerobe, and displays rods of various lengths endowed with active motility. It admits of being easily cultivated artificially, if the care is taken to fulfil the conditions necessary for the growth of anaerobes. The gelatine is liquefied in the form of roundish globular cloudinesses occurring in its substance. Surface-cultures on agar show little collections with short processes, and colonies with irregular ramifications form in like manner beneath the surface of serum. Development of a gas takes place in the cultures, the evil odour of which has given the micro-organism its name.

Bacillus œdematis maligni.—As long ago as the year 1872, Coze and Feltz found in their researches on septicæmia a micro-organism which was more fully described by Pasteur, and which obtained the name of *vibrion septique*, or *Bacillus septicus*. Koch called it the *Bacillus œdematis maligni*. The bacilli are found in the superficial layers of garden soil and in dust from the packing of the floors of rooms, as well as in various putrid matters during the process of decomposition. Van Cott found them also in unprepared musk-sacs, and from this explained the circumstance that patients are occasionally attacked by malignant œdema after injection of tincture of musk.

The best mode of carrying out an investigation is to introduce as much earth as will lie on the point of a knife beneath the skin of the abdomen of a rabbit or guinea-pig. The animal dies in from twenty-four to twenty-eight hours, and the bacilli are found in the œdematous fluid and on the surface of the organs, but not in the blood-vessels, whereas the bacilli of *anthrax* can be detected in the blood. They also differ essentially from the latter in appearance, being thinner and ending in rounded points. They unite to form curved threads, possess the power of automatic movement, depending, as R. Pfeiffer has found, upon flagella, and form central spores.

The bacilli soon lose their motility in the hanging drop, access of oxygen being fatal to them, as they are strictly anaerobic. Growth takes place either at the temperature of the room or at that of the incubator. They stain quickly in aniline dyes, but the colour is easily discharged by applying Gram's method. Consequently, the points on which reliance is to be placed in distinguishing between the bacilli of malignant œdema and those of anthrax are the form, motility, distribution in the organs, manner of staining, and relation to oxygen.

In gelatine cultures, which must be made with a regard to their strict anaerobiosis, small colonies occur, the contents of which soon liquefy, so that each forms a liquid globule in the interior of the gelatine. In the high cultures there is soon seen, as Liborius pointed out, an extensive decomposition of the nutrient medium, which is changed into a turbid fluid with simultaneous disengagement of abundant bubbles of gas (fig. 59). The addition of grape sugar to the gelatine may prove advantageous. Dull, cloudy, indistinctly-defined colonies of a shaggy appearance show themselves on agar after eight hours. They consist of a closely-woven network of finely granulated fibres, and develop lenticular gas-bubbles; indeed, this development of gas is so abundant that thick layers of agar are thrown towards the upper part of the test-tube, while a considerable quantity of a cloudy, whitish liquid condenses and gathers at the bottom. A network of bacilli forms on potato at incubating heat. The temperature most suitable for their growth lies between 37°

162   BACTERIOLOGY

and 39° C.; under 16° C. no development takes place. The spores lie at one end of the rods, and are very resistent.

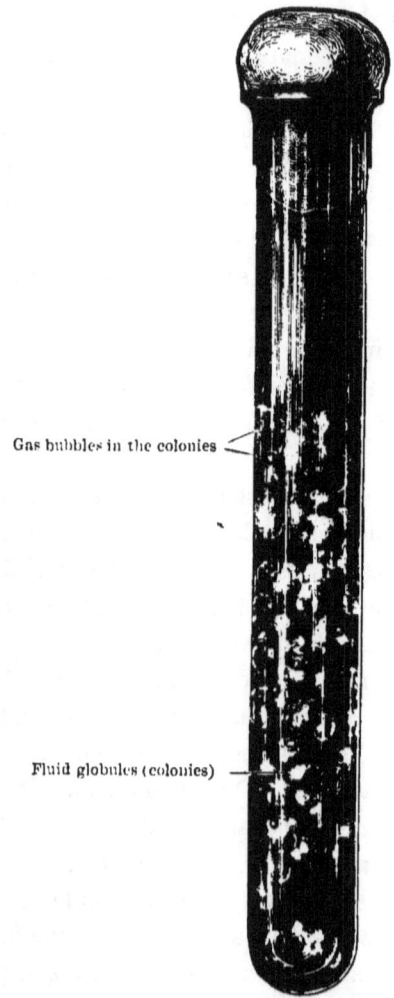

FIG. 59.—ANAEROBIC CULTURE OF THE BACILLUS OF MALIGNANT ŒDEMA IN GLYCERINE AGAR. (After Fraenkel and Pfeiffer.)

The *Bacillus spinosus* is often found in conjunction with the above, being obtained in like manner from garden mould, but, although anaerobic, it may be distinguished by the fact that its rods are not endowed with motility, and are of a different shape.

An œdema containing a reddish fluid which swarms with bacilli, and is charged with bubbles of gas, is found on making post-mortem examinations of the infected animals, and bacilli are also encountered in great numbers in the peritoneal fluid.

Cultures in bouillon retain their power of infection for a long time, even for several months.

In the case of the mouse the bacilli effect an entrance from the tissues into the circulation, probably by penetrating the walls of the blood-vessels.

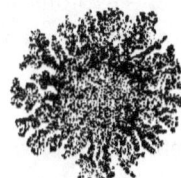   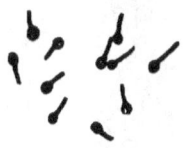

FIG. 60.—ISLET OF THE BACILLUS OF MALIGNANT ŒDEMA (*Bacillus septicus*) ON A GELATINE PLATE. (After Macé.)

FIG. 61.—TETANUS BACILLI WITH TERMINAL SPORES.

According to Penzo the bacilli of malignant œdema, notwithstanding their strict anaerobiosis, develop in ordinary cultures from which oxygen is not excluded, provided these are simultaneously inoculated with *Bacillus prodigiosus* or *Proteus vulgaris*.

The œdema bacilli break up albumen, according to Kerry, producing the ordinary putrefactive processes, and, in addition, an exceedingly poisonous oily body, which is formed by the oxidation of valerianic acid.

Bacillus of tetanus.—The *tetanus bacillus* was discovered by Nicolaier in garden mould, and in pus from the wounds of patients who had died of the disease; but Carle and Rattone had already established the fact that tetanus is communicable. The experiments proved that these bacilli are slender bristle-like rods, having circular spores at one end and possessing but little power of automatic movement, which very often range themselves in chains or clumps (fig. 61). The tetanus bacillus is rigidly anaerobic and occurs very often in conjunction with other anaerobes, so that the disease has been supposed to be due to the united action of several of these anaerobic micro-organisms. This is described as *symbiosis*.

The bacilli stand a tolerably high temperature—about 80° C.—without losing their pathogenic power, but growth takes place best at incubating heat. This peculiarity, viz. that the spores can be subjected to a high temperature without losing their vitality, enabled Kitasato to obtain pure cultures of the tetanus bacillus, since the other bacilli cultivated along with them are destroyed at a temperature of 80° C., so that it is then not difficult to procure a pure culture of the tetanus bacilli, which remain alive. A trace of pus from a patient suffering from tetanus is smeared upon serum solidified in the slanting position, or upon agar, and the cultures so made are then deposited in the incubator for some days. They are next transferred for from half an hour to an hour to a water-bath heated to 80° C., in order to kill the micro-organisms which have grown along with

FIG. 62.—ANAEROBIC GELATINE CULTURE OF THE TETANUS BACILLUS. (After Fraenkel and Pfeiffer.)

the bacilli, after which the process of cultivation on plates is proceeded with, by preference in an atmosphere of hydrogen. One or two per cent. of grape-sugar may be added to the gelatine with advantage. The plates show colonies which have a halo radiating in all directions, and liquefaction sets in slowly, being combined with the formation of gas (fig. 62). In high cultures a cloud radiating in all directions develops, which liquefies later on.

If an infection is made with a pure culture, the rods are found only on the site of inoculation and in its immediate neighbourhood.

To destroy the spores, they must be exposed for five minutes to the action of steam at 100° C.

Streptococcus septicus.—Nicolaier and Guarneri found the *Streptococcus septicus* in impure earth. It does not exhibit a chain form in all cases, and occurs in the organs of infected animals in the shape of diplococci. Gelatine is not liquefied, and little dots develop on the plate at room temperature. If mice are inoculated with impure earth they invariably die within three days; paralytic symptoms occur in the posterior extremities before death, and diplococci are found everywhere in the organs and the blood, and may block the vessels.

Bacillus anthracis.—This bacillus is also found on the surface of the ground, and its rods were seen by Pollender as long ago as 1849 in the blood of animals suffering from anthrax, although they were first thoroughly investigated by Koch. According to Pasteur, they are spread by earthworms.

They are large even rods with abruptly cut ends, arranged in chains which consist of segments of varying length (fig. 63), and are immotile, such movements of single bacteria as are now and then seen being apparently only caused by currents in the fluid. They grow at room temperature as well as in the incubator, but not below 12° or 14° C., and their highest limit of vitality is at 45° C. If the cells are frozen they again recover, according to Frisch, the capability of further development when the temperature is raised. The formation of spores can be observed with distinctness.

The cells stain readily with aniline dyes, and do not discharge their colour if Gram's method be employed, a behaviour which is of particular importance for the detection of the bacilli in the blood or organs. A minute portion of the spleen is usually taken, rubbed between two cover-glasses, and submitted to Gram's process of staining; but

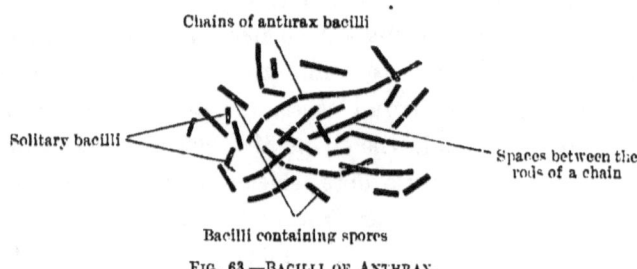

FIG. 63.—BACILLI OF ANTHRAX.

the heating must not be carried too far, otherwise the protoplasm inside the limiting membrane of the cell will undergo fine granulation. Carbolic fuchsine or carbolic methyl blue are also used for rapid staining. The cells sometimes have their ends thickened into knobs and with shallow excavations, so that an oval light spot appears between the individual cells, or, owing to the thickened nodes, they assume the figure of a bamboo rod. This latter appearance is well brought out by double staining with Bismarck brown and methyl blue.

When the anthrax bacillus is grown in bouillon, long fibres are obtained which are felted together in the form of a tress of hair, an appearance which becomes visible in twenty-four hours. On the gelatine plate little white dots appear in a day or two, and rapidly liquefy the medium, floating about on the fluid mass. The colonies are seen under the microscope to consist of irregularly arranged filaments, an appearance which is particularly marked at the border of the colony, and has been compared to the Medusa's head (fig. 64). Impression preparations of superficial colonies show the shoots and processes very distinctly. Agar plates after twenty-four hours at incubating temperature show similar figures to those on gelatine. A liquefaction beginning at the surface is seen in thrust-cultures, while

FIG. 64.—COLONY OF THE ANTHRAX BACILLUS ON A GELATINE PLATE, RESEMBLING TRESSES OF HAIR (THIRD DAY).

from the inoculated track delicate filaments push out into the gelatine (fig. 65). When liquefaction is further advanced the bacilli sink to the bottom of the funnel-shaped excavation, but no skin forms on the surface, and from the deepest parts of the liquefied area processes push into the still solid gelatine. Superficial cultures on agar form a layer which can be easily raised with the platinum needle. Serum is slowly liquefied, and a dry white coating develops on potato, with considerable formation of spores. Disinfected silk threads are often saturated with such spores, dried, and kept for experimental purposes.

168   BACTERIOLOGY

Infection experiments performed on different animals by inoculation and inhalation, as well as by admission through the digestive track after previously neutralising

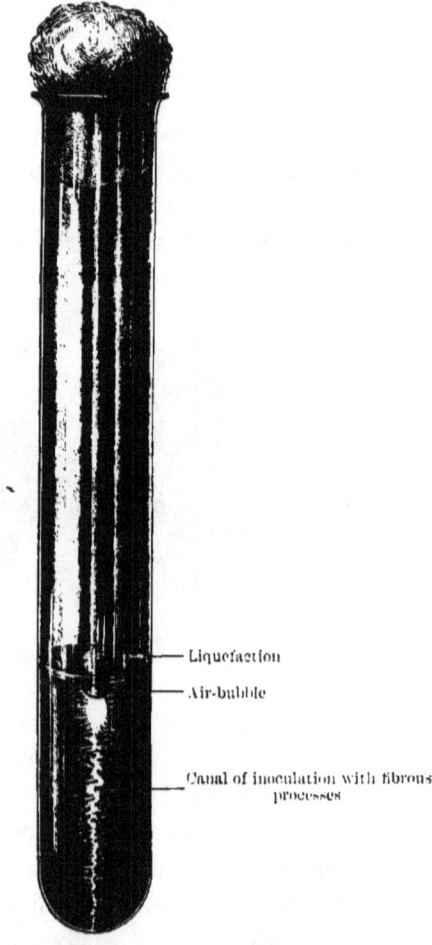

FIG. 65.—THRUST-CULTURE IN GELATINE OF THE ANTHRAX BACILLUS (FOURTH DAY).

the gastric juice, result in the death of the animal before forty-eight hours have elapsed; but in the case of frogs infection is successful only when the animals are kept at a higher than their normal temperature. If the bacilli alight upon a spot invested with epithelium, they develop there locally for some time until the epithelium is broken through and infection can take place, the result being the *malignant pustule*. According to Paltauf and Eiselsberg, the 'rag-picker's disease,' which affects persons engaged in sorting rags, especially in paper factories, is identical with anthrax of the lungs.

Owing to lack of oxygen no spores are formed by it in the body, but it is possible that they may develop on the surface of the ground where there is free access of the gas, and this renders great care necessary in disposing of the bodies of persons or animals dead of anthrax.

When it is wished to make an experiment on animals the most convenient for the purpose are white mice, which are inoculated by a pocket made under the skin near the tail. The animal dies within forty-eight hours, when the spleen is found greatly enlarged, and abundant bacilli can be detected in it as well as in the blood. A trace of the blood, or of the spleen pulp, is now used to make a gelatine plate, on which the characteristic islets are found in a few days, and a pure culture can be prepared from them.

Hankin obtained an excessively poisonous albuminoid body from cultures of anthrax, while Martin ascribes its virulence to an alkaloid.

Plasmodium malariæ.—The protozoa bearing this name, whose connexion with malaria has been established, stand in close relationship to the soil. The names associated with their discovery are those of Laveran, Marchiafava, Celli, Golgi, and Guarnieri. If a drop of blood is taken from a patient suffering from intermittent fever, at the beginning of the fit, there are found within the red corpuscles small, roundish amœboid bodies, difficult to distinguish from the corpuscular protoplasm (fig. 66). They are rendered more easily visible if the blood be smeared on a cover-glass with the edge of another, allowed to dry, and stained with watery solution of methyl blue; or, according to Celli and Guarnieri, a good method is to dissolve the methyl blue in serum or ascitic fluid, and let it run in from one edge upon a preparation of the blood, which should not be previously dried. During the attack

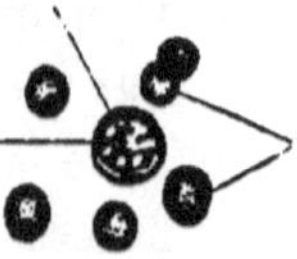

Amœboid figure in a red corpuscle

Remnant of the red corpuscle ——— Other red corpuscles

FIG. 66.—PLASMODIUM OF MALARIA IN HUMAN BLOOD, AT THE PERIOD OF APYREXIA.

of fever the plasmodia induce alterations in the corpuscle, causing a conversion of the hæmoglobin into *melanine*, so that if the blood is examined after the attack the corpuscles are found to be paler, and show in their interior clumps consisting of minute granules of black pigment (*melanæmia*). This formation of pigment advances so far that the blood corpuscles are totally destroyed, and, according to Golgi, it

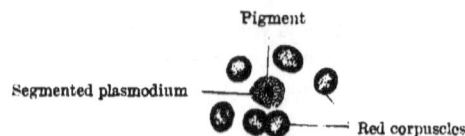

Pigment

Segmented plasmodium ———

——— Red corpuscles

FIG. 67.—PLASMODIUM OF MALARIA AT THE STAGE CORRESPONDING TO THE TIME OF ONSET OF THE FEVER.

goes on between the attacks. Multiplication of plasmodia takes place by *segmentation*, and the new plasmodia at first adhere to the edge of the blood-corpuscle, then become free and subsequently penetrate again into other corpuscles (fig. 67). According to Golgi, the process of segmentation takes two days with some plasmodia, and with others three, before the stage is reached at which the segmented portions become free, and in this way he explains the occurrence of *tertian* and *quartan* fever respectively, while *quotidian* fever is dependent on the presence of both varieties, the new spores of one kind being set free on the first day, and those of the other on the second.

In malarial cachexia there are found, besides the plasmodia, sickle-shaped bodies inside the blood-corpuscles ('*Laveran's sickles*'), and forms provided with flagella are also occasionally observed (fig. 68).

In the individual plasmodia an outer highly-refractive part which easily takes the stain (*ectoplasm*) and an inner part hard to colour (*entoplasm*) are distinguished, the latter

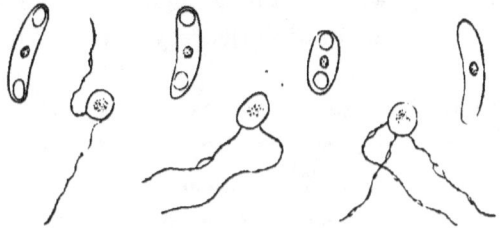

Fig. 68.—SEMILUNAR OR SICKLE-SHAPED BODIES, AND FREE BODIES PROVIDED WITH FLAGELLA. (After Jaksch.)

being enclosed by the former as by a ring. An excentrically-placed nucleus containing nucleoli is found in the entoplasm according to Mannaberg and other investigators.

Artificial cultivation has not as yet been successful. It has only been possible to preserve the plasmodia in the living leech, in the digestive canal of which, as Rosenbach found, they retain their vitality for at least forty-eight hours.

The simplest mode of staining is that of Grassi and Feletti, and is as follows :—A small drop of malarial blood is placed upon a cover-glass, which is inverted and laid upon a slide carrying a drop of aqueous solution of methyl blue or fuchsine; raising one side of the cover-glass a little and letting it fall again suffices to effect mixture of the blood and the staining fluid. By this method the nucleus appears stained darker than the rest of the body of the parasite.

Mannaberg has introduced the following method in order to study the order of events in the life-history of the malaria parasite:—The preparation after being dried in air is left for twelve to twenty-four hours in a mixture of equal parts concentrated solution of picric acid and water, to which from 3 to 5 per cent. of glacial acetic acid has been added. It is then deposited in absolute alcohol until completely decolorised, double-stained with alum hæmatoxyline solution, and differentiated in 25 per cent. hydrochloric acid in alcohol, and weak ammonia alcohol. The preparations show the red corpuscles and the protoplasm of the leucocytes unstained, but the nuclei of the leucocytes and the plasmodia strongly coloured.

Malachowski lays the cover-glass preparations in alcoho and stains them in a mixture of equal parts of a 1 per cent. solution of eosine and dilute aqueous solution of borax and methyl blue ($\frac{1}{2}$ grm. each of borax and methyl blue in 100 c.cm. water), by which method the red corpuscles show yellowish-red, the nuclei of the leucocytes violet, and the plasmodia blue.

Aldehoff recommends that the blood should be spread out on a cover-glass in the thinnest possible layer, dried in an exsiccator, and subjected for ten or twelve hours to a temperature of 120° C. in the hot-air sterilising chamber. The cover-glass is now left in a concentrated alcoholic solution of eosine for half an hour, or for two or three minutes if warmed at the same time, rinsed in distilled water, double-stained by dipping it several times in a concentrated alcoholic solution of methyl blue, and rinsed repeatedly in water. The blood must be examined immediately after being taken, to avoid mistaking blood-plates for plasmodia.

Other micro-organisms of the soil.—Besides the micro-organisms here described, a large number of others have been found which also occur in water or air and have been described under those heads, e.g. *Bacillus ramosus, Bacillus subtilis, Staphylococcus pyogenes aureus,* and *brown mould.* Karlinski also found the *typhoid bacillus* in the ground, but these bacilli soon perish on the surface of the soil, although they can maintain themselves in its deeper strata.

ANALYSIS OF PUTREFYING SUBSTANCES

Differences in putrefactive processes.—When the conditions determining the decomposition of organic substances are investigated, it is possible to recognise the presence of micro-organisms in the products which result. Decomposition must be looked on as including two distinct processes, for if the conditions necessary for the formation of its products be sought for in the mode of life of micro-organisms, it can very readily be ascertained that one variety of the process is caused by those microbes which thrive in the absence of oxygen, that is, is dependent on the vital activity of anaerobes; whereas the remaining variety is essentially a process of oxidation set up by aerobic micro-organisms.[1] In the latter, highly complex combinations are reduced to the very simplest, as happens with the manure used in agriculture; and this can take place only when oxygen has free access, and in the presence of the proper amount of water. When there is too much water the necessary quantity of oxygen cannot gain admission, and the process reverts to the first-mentioned (anaerobic) form.

During decomposition the chemical reaction very frequently changes, and there is also a formation of *ptomaines*

[1] [These processes have distinct names in German, for which there are no English equivalents, the anaerobic variety being known as *Fäulnis*, the aerobic as *Verwesung.*]—TR.

or putrefactive alkaloids possessing toxic qualities. Phosphorescence and a formation of pigment are very often observed in putrefying substances.

Bacillus fuscus limbatus.—This bacillus was discovered by Scheibenzuber at the author's Institute, in putrid eggs smelling distinctly of sulphuretted hydrogen. It consists of short rods seldom united in threads, which display an active motility and do not liquefy gelatine. Small brownish distinctly rounded clumps are at first found on the gelatine plate, and the bacilli are arranged round these in a transparent circle, forming a clear area like a collar, which is twice or three times as broad as the diameter of the original brownish islet in the centre. In thrust-cultures the growth spreads out over the surface, while along the needle-track there appear prominences like the teeth of a saw, or sometimes bowed outwards. The medium shows a discoloration round the puncture in the form of a bag with the convexity downwards ànd the constricted opening above. This appearance, which is characteristic, shows it to be an organism which, like *Bacillus fluorescens non-liquefaciens* and *Bacillus melochloros*, imparts its colouring matter to the nutrient medium. On agar and potato a brown growth appears. It thrives at the temperature of an ordinary room, as well as at that of the incubator.

Proteus.—The varieties of *Proteus* described by Hauser belong to the pathogenic micro-organisms living in putrid substances. The kinds distinguished are *Proteus vulgaris*, *Proteus mirabilis*, and *Proteus Zenkeri*, organisms which set up putrid decomposition. If animals are inoculated subcutaneously with small quantities of cultures of *Proteus* toxic phenomena are evoked, and the animals, whether rabbits or guinea-pigs, die in a comparatively short time with the symptoms of a severe peritonitis or enteritis. The toxic action originates in the decomposition of albuminoid bodies, from which poisonous substances are split off.

The **Proteus vulgaris or Bacillus figurans** consists of small, curved, very actively motile rods, which also occur united in groups; but sometimes the original form alters, so that cocci result. Brown colonies first appear on the gelatine plate, the margins of which carry tufts of processes, and when these coil after the manner of tendrils, the result is a figure (*wandering islet*) which is easily secured by means of an impression preparation and rendered more distinct by staining. When examined under a moderately high power such a preparation shows the elements and the arrangements formed by their union. A deposit of bacilli arranged in combinations is found in the liquefied areas in the interior of the gelatine. Thrust-cultures show a very rapid liquefaction, the micro-organisms sinking to the bottom. A coating forms upon agar, and a dirty film upon potato. Serum is liquefied and undergoes very speedy putrefaction, as is also the case with masses of meat.

Proteus vulgaris has the power of causing milk to turn sour, according to Kühn, or bitter, according to Krüger. It has also the power, as shown by J. Schnitzler, of converting urea into ammonium carbonate.

The **Proteus Zenkeri** does not liquefy gelatine and grows also when air is excluded. Its colonies can be readily raised from the gelatine plate.

Proteus mirabilis liquefies gelatine very slowly. Of the remaining varieties of Proteus, the **Proteus hominis** and **Proteus capsulatus** must be mentioned.

Bacillus saprogenes.—Rosenbach found bacilli in foul-smelling secretions, which, when cultivated, exhale a strong putrid odour, and cause a stinking putrefaction of flesh. He distinguishes three varieties of *Bacillus saprogenes*. The first kind, which he isolated from the white plugs met with on the wall of the pharynx, shows on agar a yellowish-grey, opaque, pap-like streak (*Bacillus saprogenes I.*). The second kind, which he found in the foul-smelling perspiration on the feet, exhibits numerous fine droplets upon agar, which spread out and form a film as clear as water (*Bacillus saprogenes II.*). The third kind was found by him in septic gangrenous pus. It develops a broad and nearly fluid film on agar (*Bacillus saprogenes III.*).

Spirillum concentricum was found by Kitasato in putrid ox-blood, and shows elements twisted into a screw shape and furnished with flagella, which impart to them a lively motility. Roundish, sharply-defined discs occur on the gelatine plate, and display a concentric stratification con-

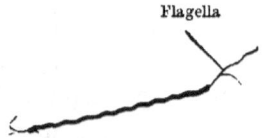

FIG. 69.—SPIRILLUM RUBRUM WITH FLAGELLA. (After Löffler.)

sisting of alternate transparent and opaque rings. The gelatine is not liquefied. In thrust-cultures the microbe grows mostly on the surface. The layer formed on agar is firmly adherent; on glycerine-agar the concentric arrangement is distinctly apparent, while on potato no growth takes place.

Spirillum rubrum.—This was discovered by Von Esmarch in a mouse which had died of *mouse septicæmia*. The spirilla possess an active motility, and are provided with flagella, according to Löffler (fig. 69). Von Esmarch grew them first in roll-cultures on gelatine. The colonies are grey at the commencement, afterwards red; they grow with extraordinary slowness, and do not liquefy the gelatine. A special point is that thrust-cultures show after some time

SPIRILLUM RUBRUM

a wine-red colour, which, however, does not extend to the surface, and hence the pigmentation does not appear to depend on the action of oxygen. Light exercises an inhibitory influence on its production. On agar a white layer first forms, which later becomes rose-red. Only small red islets appear on potato; in bouillon growth occurs, as also in the water of condensation from serum. The microbes grow both at room temperature and that of the incubator.

CHAPTER VIII

MICRO-ORGANISMS IN ARTICLES OF DIET

Methods of examining different foods.—As in examining the micro-organisms in water and in the soil, so also articles of diet are tested for the bacteria which they contain by means of the process of plate-culture. Many such articles are impregnated with bacteria, owing to contamination with adherent particles of earth or germs from the air; whilst probably insects, particularly flies, play an important part in the transmission of the microbes to them. Many foodstuffs, again, form a very favourable nidus for some of the bacteria—for example, milk, meat, and soup for the bacilli of *typhoid fever* and *cholera*—while others, such as the milk and flesh of animals suffering from *tubercular disease* (*Perlsucht*), are directly infectious, owing to the bacteria in them.

In the case of vegetables aqueous infusions are made, from which samples are taken for microscopic examination. Cheese is examined by rubbing up a little of it with water, and using minute particles of the mass to make plate-cultures; milk is treated like other liquids, and butter and similar substances are submitted to examination either with or without the addition of fluid, according to their consistence.

When it is not possible to prepare plate-cultures immediately, a sample of the substance to be examined is taken with the sterilised platinum wire, and a thrust-culture in gelatine made therewith, from which further investigations can be carried on. Still, this procedure comes far behind the plate method in certainty, and the latter should always be employed where it is at all possible.

EXAMINATION OF MILK AND ITS PRODUCTS

Methods.—Milk may be contaminated either during milking or in the subsequent manipulation, and may exhale an odour owing to the substances so acquired, or become slimy or stringy, or take on a bitter or an acid taste, and its colour may turn blue or red. Moreover, various pathogenic micro-organisms may be imparted to milk from the animal furnishing it. To examine, Arens recommends that a loopful of milk be diluted with another of distilled water upon a cover-glass, dried, and fixed by heat, which should not be too strong. The cover-glass so treated is now brought into chloroform methyl blue, prepared by mixing twelve to fifteen drops of saturated alcoholic solution of methyl blue with three or four cubic centimeters of chloroform. It is then waved to and fro for from four to six minutes, so as to allow the chloroform to evaporate, and the adherent methyl blue is rinsed away. In fresh milk and in cream the bacteria alone are dark blue; in curdled milk the flakes of caseine are also stained, but only a pale blue.

Another method of examining milk bacteriologically consists in treating a drop placed upon a cover-glass with two or three drops of a one per cent. solution of sodium carbonate, the fluids being mixed as thoroughly as possible with the platinum needle, after which the cover-glass is carefully heated over a small flame until evaporation is complete. The result of this process is saponification of the fat, so that the cover-glass appears overlaid with a thin film of soap, and the preparation so treated can be submitted to the ordinary staining processes.

Bacillus lacticus (Bacillus acidi lactici).—Pasteur pointed out that the conversion of sugar into lactic acid (lactic acid fermentation) is due to the action of micro-organisms. Hueppe has investigated this fermentation more thoroughly, and has described a microbe possessing in an eminent degree the property of curdling milk. It consists of short, plump, immotile rods, which usually occur in pairs but are occasionally united in longer chains. Growth can take place in the absence of oxygen, as well as when it is present. On the gelatine plate little white points appear, having a surface like porcelain, which do not liquefy the medium; they show under the microscope a yellowish coloration in the centre, which becomes more transparent and paler towards the margin. Growth appears along the needle-track in thrust-cultures; a coating of considerable extent soon forms on the surface, and crystals of salt separate out from the nutrient medium. A whitish layer develops on agar, and a thick brown dirty deposit on potatoes.

If a small quantity of a pure cultivation of *Bacillus lacticus* be added to milk, formation of acid occurs, with curdling of the caseine, the milk-sugar being split up into carbonic and lactic acids. This action on milk seems to be capable of taking place only when air has access, although the growth of the micro-organism is independent of it.

Micrococcus acidi lactici.—The micrococcus of this name described by Marpmann closely approximates in its action to the *Bacillus lacticus*. It consists of non-motile cocci which occur singly or in pairs. In twenty-four hours yellowish-white punctiform colonies appear on the gelatine plate, while in thrust-cultures a superficial layer develops which is thicker in the centre than at the edge, and is rather slow in growing. Reddening of the milk takes place in twelve hours, and in twenty-four coagulation and lactic acid fermentation.

Marpmann has also found the **Sphærococcus acidi lactici** in milk. It possesses properties similar to those of the *Micrococcus acidi lactici*.

Clostridium butyricum, or Bacillus amylobacter.—This is found in old milk, and is also very widely distributed elsewhere in nature. It was thoroughly studied by Prazmovsky. It is very often met with in cheese as well as in putrid vegetable infusions, and is, as discovered by Nothnagel, a very frequent inhabitant of fæces. The rods are actively motile and strictly anaerobic, and develope in gelatine and agar gases which smell of butyric acid. They stain blue or dark violet in aqueous solution of iodine, from which it has been concluded that the protoplasm must contain *granulose*—hence the name *Bacillus amylobacter*. Butyric acid is formed from starch, dextrine, and salts of lactic acid with simultaneous development of hydrogen and carbon dioxide gas; while masses of caseine and other coagulated albuminoid bodies are liquefied.

Micrococcus acidi lactici liquefaciens.—Krüger has discovered this micro-organism in caseous butter and in cheese. It consists of small oval immotile cocci, often arranged in groups of four, and forms small colonies on the gelatine plate, which speedily liquefy that medium. In thrust-cultures there develops on the surface of the funnel-shaped area of liquefaction a white pellicle, which later sinks to the bottom. Milk curdles at 25° to 35° C., with formation of lactic acid and separation of a clear serous layer on the surface, while after a longer time a smell like that of paste is given off. The micro-organism grows whether air is excluded or not.

Acidity and bitterness in milk may be also occasioned, according to Kühn, by the *Proteus vulgaris*, which causes a thickening of the milk.

Oidium lactis.—In sour milk, and constantly in butter,

there is found the *Oidium lactis*, described with considerable completeness by Grawitz, which forms a white fur like a mould. The fibres of the mycelium mount upwards, become segmented, and support rows of cylindrical gonidia. The oidium grows on gelatine without liquefying it, and forms a white, long-haired fur on the suface of the plate at 20° C., diffusing at the same time an odour as of sour milk. The thrust-culture appears permeated with fibres. Gelatine which has an acid reaction is liquefied. Whitish stars develop on agar at 30° C., which coalesce and cover the surface. Growth takes place also on serum, and on milk a thick coating of mould forms. (See fig. 5.)

Bacillus butyricus.—According to Hueppe the butyric acid fermentation is caused by this bacillus, which consists of short curved rods endowed with a lively motility. They have the power of curdling milk, and originate the above-named variety of fermentation. The curdled caseine is dissolved and converted into peptone with simultaneous splitting off of ammonia and other products of division, the milk acquiring at the same time a bitter taste. Gelatine is very rapidly liquefied. Yellowish colonies appear on plates; in thrust-cultures a greyish-white skin develops, and also a yellow coloration of the liquefied gelatine, while on agar a yellow layer is formed, and a wrinkled, dirty coating upon potato. Centrally-situated spores develop at incubation temperature.

A genuine *Bacillus butyricus* is, according to Botkin, to be met with in all milk, and by the liberation of free butyric acid causes rapid curdling with abundant evolution of gas. To isolate it, the milk is sterilised for half an hour in the steam steriliser and, after being sealed up air-tight, is deposited in the incubator. It is completely decomposed in two weeks, and is then used to make an anaerobic plate-culture, best done, according to Botkin, by employment of a bell into which hydrogen is conducted in the presence of pyrogallol, the apparatus being luted with paraffin. The same observer recommends sugar-agar as a nutrient medium. After continuing in the incubator for two days, the colonies have made such progress that they can easily be transplanted as pure growths to high cultures in agar. Round colonies form, which give the impression of a felt consisting of closely-intertwined filaments, and are provided with a considerable number of processes radiating out from all sides. Gelatine is very rapidly liquefied. On potato there develops at the depth of a millimeter an abundant growth of bacilli, which loosen the substance of the medium to a marked degree, and exhale a peculiar odour resembling that of alcohol.

Bacillus butyri viscosus.—Lafar constantly found short rods in butter, having rounded ends and not seldom forming chains. Thrust-cultures spread but little over the surface of the gelatine, but along the needle-track there occur figures resembling the scales of fish, in lumps, or often in clusters, and which do not liquefy the medium. Small white dots appear on the surface of the gelatine plate, which grow decidedly more in height than in width, gradually melt into the form of small lenticular discs, and cover the whole of the gelatine, which remains solid, with a slimy film about half a millimeter in thickness. On potato a fine moist shining layer develops, resembling that formed by typhoid bacilli. Growth takes place whether air is excluded or present.

The **Bacillus butyri fluorescens**, isolated from butter by Lafar, seems to be identical with the *Bacillus melochloros* obtained from air by F. Winkler and Von Schrötter in the author's Institute. (See p. 118.)

Spirillum tyrogenum.—Deneke cultivated from old cheese a variety of comma-bacillus which has strongly curved elements endowed with rapid automatic motion. On the gelatine plate small punctiform colonies occur, which liquefy the gelatine and acquire a yellow colour. They resemble at first a culture of the *cholera vibrio*, but are distinguished from it by the speedier liquefaction of the gelatine, and by the yellowish tint of the colonies. Liquefaction is, however, slower than in the case of the *Vibrio proteus*. In thrust-cultures the gelatine is liquefied along the whole length of the inoculated track, and in a few days the bacteria sink to the bottom and lie there in a loose coil. The bubble of air in the upper part is larger than in cholera cultivations (see fig. 53). A thin yellowish coating appears upon agar, a yellowish film upon potato, and serum and plovers'-egg albumen are liquefied. If injections of the cultures be made into the intestinal canal of guinea-pigs, with the precautions specified by Koch in the case of cholera bacilli, the animals are destroyed.

Bacillus lactis viscosus.—This very wide-spread spoiler of milk was first found by Adametz in the water of brooks in the neighbourhood of Vienna. It forms short rodlets resembling cocci, which possess a thick, highly-refractive capsule. On gelatine plates round disc-like colonies occur, sometimes with concentric rings, and on them there forms at a favourable temperature (16° to 20° C.) a broad, irregularly serrated hem of a transparent, horny appearance. Surface-cultures upon agar show narrow whitish stripes, the border of which is at first smooth, but later finely serrated. Sterilised milk becomes in four to six weeks viscous like honey, so that it can be drawn out into long strings (*slimy fermentation*); in milk which has not been sterilised the cream alone becomes stringy or slimy, and such cream yields a white, greasy butter which becomes quickly spoilt owing to the presence in copious numbers of the butyric acid bacillus, from which fact Adametz concludes that this bacillus prepares the soil for that of butyric acid. The stringy substance comes from the material of which the envelopes of the bacteria are formed, and is probably metamorphosed cellulose, as in the case of *Bacillus mesentericus vulgatus*, which also has the power of setting up a slimy fermentation.

Such fermentation is also originated, as Löffler found, by the **Bacillus lactis pituitosi**, which consists of thick curved rods very rapidly breaking up into segments like cocci. The colonies on gelatine plates are radially striped, and the gelatine is not liquefied. On potato a greyish-white coating develops.

Bacillus actinobacter.—Duclaux very frequently found in milk fine non-motile rods, partly isolated, partly arranged in pairs, but always furnished with jelly-like capsules, and which rendered the milk gelatinous and viscid. The capsules are retained in artificial cultures in glycerine solutions, whereas in growths in bouillon and solutions of sugar they are lost. Development can take place either in the presence or absence of oxygen.

Bacillus fœtidus lactis.—Jensen isolated a bacillus from milk which imparts to it and its products a nauseous, sweetish, putrid odour and taste. The bacillus shows short thick rods, with rounded ends and active motility. Gelatine is not liquefied. Superficial colonies with the lustre of mother-of-pearl occur on plates. In thrust-cultures a slimy growth appears on the surface, and an abundant development takes place along the track of inoculation. On agar there is a formation of large lenticular bubbles of gas.

Bacillus cyanogenus.—The cause for the development of a blue colour in milk is to be sought for in the *Bacillus cyanogenus*, described by Ehrenberg as *Bacillus syncyanus*, and which was studied by Hueppe and Neelsen. The bacillus of blue milk consists of small, actively-motile rods provided with abundant flagella, and which do not liquefy gelatine. Upon the plate there form superficial round colonies, which cause a dark coloration of the surrounding gelatine. Thrust-cultivations grow in the form of the *nail-culture*, the head of which consists of a thick white deposit. The gelatine is stained greyish-blue, and finally assumes a dark tint. On agar a superficial coat forms, and the medium becomes discoloured. On potatoes there occurs at first a thick, yellow, dirty layer round the inoculated spot, and in older cultures a blue coloration of the medium. The bacillus multiplies by formation of spores occupying a terminal position. The blue pigment is changed to a brilliant red by alkalies, but is restored by the action of acids. According to Gessard, the substance from which it is derived is lactic acid.

Bacterium lactis erythrogenes.—Hueppe and Grotenfeldt found this microbe in red milk, and also in the fæces of children. It consists of short, oscillating rods, the growth of which liquefies gelatine. The colonies are of a yellow colour when first seen on the plate, but after liquefaction they become rose-red. On thrust-cultures there appears a whitish, or later, yellowish layer, which finally acquires the rose-red tint; when liquefaction has taken place in ten or twelve days the fluid is rose-coloured, and the coloration extends also into the substance of the remaining solid gelatine. A yellowish deposit occurs on agar, and soon changes to yellowish-red. On the surface of potato a golden-yellow colour appears in six to eight days at incubating temperature, whereas at ordinary room temperature the colonies are yellowish-red. All cultures exhale an unpleasant sweet smell.

If a small quantity of pure culture be added to neutral or feebly alkaline milk, the caseine is separated out and peptonised, the layer of cream on the surface and the precipitated caseine at the bottom remaining white while between them lies the rose-coloured whey. The formation of pigment takes place in the dark as well as by daylight. According to Hueppe, it appears to result from metabolism, that is, to be a coloured ptomaine, and is seemingly independent of other influences and conditions.

Red coloration of milk may also be due to the *Bacillus prodigiosus*, which, however, causes only a very slow separation out of the caseine, and does not originate any further changes.

Sarcina rosea.—Menge has found that the agent producing the red tint may be a sarcina to which he has given the above name. In does not, however, appear to be identical with the micro-organism of the same name described by Schröter (see p. 109). In two days small translucent, perfectly circular colonies form on the gelatine plate, and soon assume the figure of a rosette having a red nodule in the centre surrounded by concentric rings. In the thrust-culture the surface is covered with a thin rose-red coat with jagged edges, whilst the needle-track remains colourless. Liquefaction takes place very late. On agar a coherent growth develops which is white at first but becomes coloured on the third day; in the incubator, however, pigmentation fails altogether. On potato which has been rendered alkaline the sarcina grows excellently. The pigment forms in milk independently of light.

The Micrococcus of bovine mastitis.—In different diseases originated in the lacteal glands of animals by the action of micro-organisms the latter can be found in the milk also, and amongst these is the *micrococcus of mastitis* in the cow described by Kitt, which is met with in the partly milky and partly purulent contents of the udder of cows suffering from this complaint. On the gelatine plate small punctiform colonies of the size of a pin's head develop, and a thrust in the same medium produces a nail-culture without liquefaction. A wax-like layer appears on potato.

Other pathogenic bacteria in milk.—In the milk of animals suffering from *tuberculosis* bacilli may be found, according to Bollinger, even before the udders become diseased. *Typhoid bacilli* may also find their way into milk along with water, and may develop abundantly there. The bacteria of *cholera*, too, are similarly capable of copious multiplication in milk without perceptible alteration, but they are destroyed when lactic acid fermentation sets in. In milk from a cow suffering from inflamed udder, the *Staphylococcus pyogenes aureus* has been detected by Krüger.

Saccharomyces ruber.—Demme has found a yeast occurring in milk and cheese which produces in the latter red punctiform accumulations of pigment, and in the milk a red sediment. This yeast he has described as *Saccharomyces ruber*. Gelatine is not liquefied by it, and it develops on the plate colonies of the size of a grain of millet-seed, which show no red colour until a week has elapsed. In thrust-cultures the growth is chiefly on the surface. Discs of potato are covered with a coating of the colour of raspberries in from eight to twelve days.

Bacillus Caucasicus (Dispora Caucasica, or Kephir bacillus). —The name of *kephir grains* is given to masses of micro-organisms in a zoogloea, which are used in the Caucasian districts for the preparation of the drink known as 'kephir' from milk. In this process the caseine passes into solution, while the milk-sugar is affected after the manner of a lactic acid and alcohol fermentation (Adametz). The granules contain yeast and bacteria capable of peptonising caseine, viz. the *Bacillus acidi lactici*, *Bacillus butyricus*, *Bacillus subtilis*, and in addition to these a micro-organism to which Kern ascribes the principal *rôle*, and which bears the name *Bacillus caucasicus*. This consists of short cylindrical rods which are destitute of movement so long as they are in zoogloea, but when isolated possess an extraordinarily lively motility. The name *Dispora* was assigned to the micro-organism from the fact that the protoplasm is withdrawn to the ends, giving the rod the appearance of being divided into two spores. By its vital action milk-sugar is changed into glucose, upon which the yeast can then begin to work.

EXAMINATION OF OTHER ARTICLES OF DIET

Bacillus megaterium.—This bacillus was found by De Bary on boiled cabbage-leaves. It consists of long, thick, somewhat curved rods with rounded angles, frequently connected in chains, and possessing slightly granular cell-contents. The motion of the rods recalls the amoeboid movements of cellular protoplasm. During involution the shape of the rods loses its distinctness and malformations occur, which, however, regain their normal outline on

Chain with links containing spores
FIG. 70.—BACILLUS MEGATERIUM WITH SPORES.

FIG. 71.—ISLET OF BACILLUS MEGATERIUM ON A GELATINE PLATE.

transference to fresh nutrient substances. Their sporulation is particularly suitable for purposes of study, and the spores also stain easily (fig. 70). On the gelatine plate there form minute roundish colonies consisting of undulating filaments, which liquefy the medium slowly, and possess a uniform or semilunar outline (fig. 71). A funnel-shaped liquefaction appears in thrust-cultures, in which the bacteria sink to the bottom (fig. 72). On agar grey deposits develop, which adhere firmly to the mass, and on potato there grows a dirty whitish-grey layer containing many involution forms.

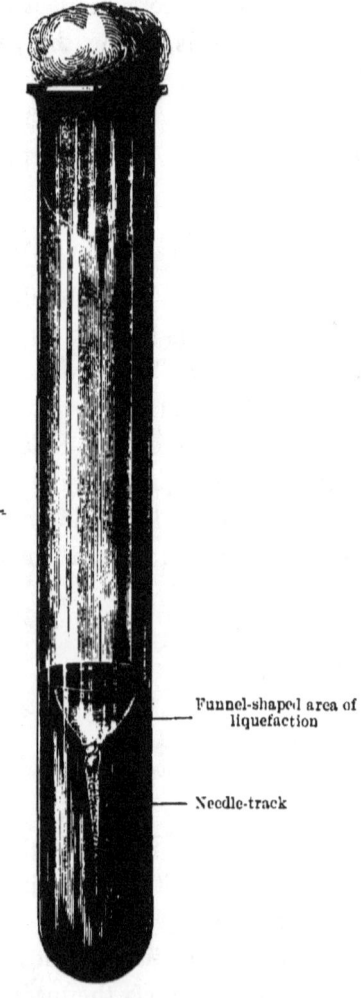

Fig. 72.—Thrust-Culture in Gelatine of the Bacillus megaterium.

Besides the above, the *Bacillus prodigiosus, Bacillus subtilis, Bacillus amylobacter,* and *Bacillus butyricus* are met with on plants. F. Winkler and Von Schrötter found the *Staphylococcus pyogenes aureus* in rotten portions of apples, and the *Bacillus melochloros* in the excreta of caterpillars in worm-eaten specimens of the same fruit.

Bacillus aceti.—As early as 1864 Pasteur brought forward proof that the oxidation of alcohol and its alteration into acetic acid are connected with the *vinegar ferment*, which could only develop in the presence of oxygen. This ferment consists of short thick rods often uniting in curved chains, and forming a mass of zoogloea which is thick and glutinous, the [so-called] 'mother of vinegar.'

Macé found another bacterium of vinegar, the zoogloea mass of which is thick, white or slightly rose-tinted, never wrinkled, and feels almost like cartilage. Numerous rods, sometimes isolated, sometimes joined in pairs or in threes, lie in a colourless interstitial substance; they are without movement in the membrane, but when free in fluids possess a slow motility. Gelatine is not liquefied, and a thick, undulating, rather hard layer develops on its surface. A less hard, but smooth and non-undulating, layer forms upon agar.

The bacteria of vinegar are very widely diffused through nature, according to Duclaux, who assigns an important part in transmission of the ferment to a kind of fly, the *Musca cellaris*, which is met with everywhere.

Bacillus indigogenus was found by Alvarez in infusion of indigo leaves. It consists of short motile rods surrounded by an envelope, and has the property of causing the appearance of blue indigo in decoctions of the leaves of the Indigofera. Surface cultures on agar cause cleavage of the medium, with development of gas.

Pediococcus cerevisiæ.—Balcke found the *Pediococcus cerevisiæ* in beer, in breweries, and in the washings from them. It forms diplococci and tetracocci, and shows colonies on plate-cultures which are at first colourless and later brownish, and which do not liquefy the gelatine. In thrust-cultures a white, leaf-like layer is developed. An iridescent greyish-white coating occurs upon agar, while upon potato the growth is scanty. Some formation of lactic acid accompanies its development.

Sarcina cerevisiæ.—A number of sarcinæ derived in part from the air are found in beer. Lintner encountered a variety in soured beer which, according to Adametz, occurs also in water, and which sets up lactic acid fermentation. Gelatine is not liquefied. The islets formed on plate-cultures are round, colourless, and smooth-edged, and a delicate fluorescent film develops gradually over the gelatine. Thrust-cultures show a smooth white coating, and on potato light granular islets form.

Micrococcus viscosus.—A peculiar disease of wine and beer, called by the French *la graisse*, which consists in a clouding and inspissation of the liquor, so that it becomes stringy like white of egg, depends, according to Pasteur, upon the vital activity of the *Micrococcus viscosus*. The cells occur singly, or more usually arranged in diplococci or streptococci. Artificial cultivation succeeds best in solutions containing sugar.

Bacillus viscosus cerevisiæ.—In viscous beer Van Laer constantly found slender rods with no tendency to unite in groups. Gelatine is liquefied. Upon plates there develop round or oval colonies with uneven edges, which project somewhat above the surface. In thrust-cultures a white, irregularly-edged deposit is observed, while the track of inoculation also shows an abundant development of colonies. Growth on agar is very rapid, with the formation of a broad white slimy layer. These bacteria also occur in milk and render it slimy. White, watery, and very tenacious colonies develop on potato, and give off an odour of putrid fish.

Bacillus viscosus sacchari.—Kramer assigns long rods with rounded corners, which are destitute of motility and frequently arranged in chains, as the cause of solutions of sugar becoming slimy. Gelatine is liquefied, a whitish film develops upon agar, and a firm mass upon potato. No growth takes place on acid media.

Moulds on articles of food.—The moulds find abundant opportunity for development and propagation upon vegetable foods, and the three varieties, *Penicillium, Aspergillus*, and *Mucor*, are all represented. *Penicillium glaucum* has already been described under the 'Bacteriological Analysis of Air' (see p. 103).

The various kinds of **Aspergillus** flourish on bread and candied fruits. To obtain the **Aspergillus niger**, bread pap is prepared in an Erlenmeyer's flask. After some time stout fructifying hyphæ are found, not ramified at the end like Penicillium, but swollen so as to resemble clubs. Upon these are arranged the sterigmata, at the upper ends of which the spores become segmented off and form aggregations which take the form of rounded, bulging black swellings. The entire flask soon becomes filled with fibres and grey points. **Aspergillus albus** and **Aspergillus glaucus** are similar, but grow better at incubating temperature (see fig. 3).

The **Aspergillus flavescens** is distinguished by its well-marked fructifications and the greenish colour of the cultures. **Aspergillus fumigatus** bears very fine fructifications, and forms an ash-grey fur. Both grow luxuriantly on bread at incubation temperature. On gelatine plates filaments appear which spread rapidly into the surrounding parts with liquefaction of the medium.

Of the **Mucorineæ**, the *Mucor mucedo, Mucor corymbifer*,

and *Mucor rhizopodiformis* are found upon food-stuffs, particularly bread. They possess a branching mycelium, with spore-bearers which resemble flexible tubes, are unsegmented, and stand up vertically from the mycelium. At the upper ends of these are the swollen *sporangia*, by bursting of which the spores are set free. In addition to this process, a kind of conjugation also takes place, in which two cells (*zygospores*), developed from the mycelium, coalesce with one another and form spores.

The **Mucor mucedo** is one of the commonest moulds, and is frequently found in animal excreta, especially in horse-droppings. It grows on acid media, and is not pathogenic. A dense fur with black fructifications as large as poppy-seeds appears on the gelatine plate.

The **Mucor rhizopodiformis** was described by Lichtheim. It grows very luxuriantly on bread gelatine, which it liquefies. The fur is white and bears black fructification heads. The culture on bread is distinguished by the development of an aromatic odour.

Mucor corymbifer was likewise described by Lichtheim. It forms a dense snow-white fur upon bread, resembling teased-out cotton-wool.

The **Mucor ramosus**, described by Lindt, grows very well upon agar made with bread infusion, and upon potato. The fur is at first white, but soon assumes a greyish-brown tint.

Both *Mucor rhizopodiformis* and *Mucor corymbifer* are pathogenic. Intravenous injection of fluid containing their spores causes a fatal disease in rabbits, in which the organs are always attacked in the following order : viz. kidneys, intestine, mesenteric glands, spleen. *Mucor ramosus* acts quickest, an acute hæmorrhagic disorder being originated by its injection.

The *Aspergillus fumigatus* and *Aspergillus flavescens* in

like manner exhibit pathogenic properties. Lichtheim found that a peculiar disturbance of equilibrium occurred in rabbits and dogs after intravenous injection, and after death, which took place in twenty-four hours, the germinations were found especially in the myocardium and kidneys.[1]

[1] [The three varieties of *Aspergillus* may grow in various parts of the body in man, notably the external ear, in which they produce the disease known as *otomycosis*. One or other form has also been found growing in the lungs, on the nasal mucous membrane, and on the cornea. Such occurrences are, however, very rare.]—Tr.

CHAPTER IX

BACTERIOLOGICAL EXAMINATION OF PUS

Properties and composition of pus.—Pus shows an alkaline reaction and possesses a high specific gravity, consequently it furnishes an excellent nutrient medium for the most widely differing micro-organisms, and that whether it is derived from exudations or from the surface of wounds, or is formed in some other inflammation of tissue.

Furthermore, pus may be coloured, the tint being greenish or brown-red, and it exhales a peculiar odour. In it are found white and red blood corpuscles, blood pigment, and crystals of hæmatoidin, epithelial cells, drops of fat, fungi, micrococci, and bacilli.

Suppuration may also be excited by bringing certain of the micro-organisms contained in pus into contact with the tissues, when their vital activity will start the process. Amongst these microbes is included more especially the *Staphylococcus pyogenes*, which, as we have already remarked in the chapter on the 'Bacteriological Analysis of Air' (see p. 110), is very widely distributed in nature. Rosenbach also found the *Bacillus saprogenes III* (see p. 176) in pus.

Actinomyces.—A fungus has been detected comparatively recently in almost all organs, in certain cases of chronic inflammation combined with suppuration, which has been named *Actinomyces* or *ray-fungus*. It had been discovered by Langenbeck as long ago as 1845, but it was not until 1878 that it was more thoroughly described by Israel. The disease caused by it is known as *actinomycosis*. E. Ullmann states, however, that the actual suppuration in this disorder is due to the specific organisms of pus, viz. the staphylococci and streptococci, the actinomyces fungus being, according to him, merely an accidental occurrence in the pus.

The fungus forms minute sulphur-yellow nodules of the size of poppy-seeds, which, if a cover-glass is lightly pressed upon them and they are examined even with a very low power, show compact globules having a clustered arrangement. They were first discovered by Bollinger in cattle, and admit of being very easily transmitted to human beings. Under a higher power they are seen to be made up of filaments resembling hyphæ, which for the most part radiate out from a central point, each ray being thickened into an elongated club-shaped enlargement at the periphery.

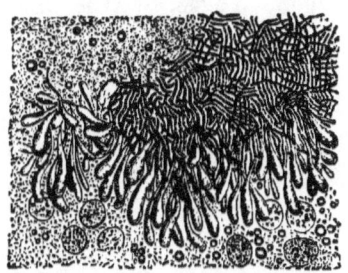

FIG. 73.—AN ACTINOMYCES GRANULE. UNSTAINED PREPARATION. (After Jaksch.)

R. Paltauf describes the actinomyces fungus as a bacterium, regarding the club-shaped figures as degenerative forms; Israel and Wolff count it amongst the *polymorphic* fission fungi, on the grounds that thread-like figures and groups of globular bodies are visible in addition to the radiating forms; Petroff and Flormann found that the fibres are made up of granules alternating with small roundish gaps; and Rabe is of opinion that a certain alga, *Cladothrix canis*, presents a very great similarity to actinomyces. The fungus takes aniline stains when they are allowed to act on it for some time, and does not discharge them during the application of

Gram's method; but the simplest mode of recognising them in pus consists in examining the little yellow globular bodies without staining (figs. 73 and 74).

Cultivations on glycerine agar show, according to Protopopoff and Hammer, a mass of closely-packed miliary nodules of the size of hemp-seed at largest, which have a yellow colour and adhere very firmly to the medium. The growth on potato is similar, except that the culture looks quite dry. In broth miliary nodules develop after a short time, and may attain the size of a hazel-nut. Gelatine is slowly liquefied. Growth takes place also in the absence of oxygen, in eggs, by Hueppe's method.

FIG. 74.—ACTINOMYCES, STAINED BY GRAM'S METHOD. (After Jaksch.)

F. Winkler has cultivated actinomyces on plover's egg albumen in the author's Institute. Formation of sulphur-yellow colonies as large as poppy-seeds takes place all over the surface of the medium, which is, however, not liquefied.

Bacillus pyocyaneus.—This bacillus is the cause of the grey or blue colour sometimes seen in pus and in pieces of dressing saturated with it.

The **Bacillus pyocyaneus** *a*, described by Gessard, consists of small slender rods provided with a flagellum, by means of which they move swiftly. Upon gelatine plates roundish islets of a yellow colour appear in the substance of the medium, to the whole of which they impart a greenish tint in about two days, causing at the same time slow liquefaction. In thrust-cultures the gelatine liquefies along the needle-track, the fluid mass assuming a greenish colour, while the part still solid displays a green fluorescence. A layer forms upon agar which shows a fluorescence, at first greenish, but subsequently dark green in tint; while on potato the deposit appears coloured brown, turning red when treated with acids, blue with ammonia. According to Gessard, the generation of pigment depends on the nutrient medium. The green colouring matter formed, which is soluble in chloroform, has received the name of *pyocyanin*.

Rabbits die when injected into the peritoneal cavity with pure cultures of this micro-organism or with pyocyanin.

The **Bacillus pyocyaneus** β, described by Ernst, is not pathogenic, but is commonly met with in the company of the foregoing, and when isolated causes the blue coloration of pus. It liquefies gelatine much more speedily than the last, and shows a brown tint on potato which changes to grey on touching with the platinum wire (the *chameleon phenomenon*). The reaction of the potato culture, the more rapid liquefaction, and the fact that injections produce no result, all serve to distinguish it from *Bacillus pyocyaneus a.*

FIG. 75.—ISOLATED ELEMENTS OF STAPHYLOCOCCUS CEREUS FLAVUS.

With reference to its physiological relations, *Bacillus pyocyaneus* produces coagulation of milk, rapidly peptonises albumen, and causes vigorous inversion of sugar with fermentation.

Staphylococcus cereus.—Passet found in pus the **Staphylococcus cereus albus** and **Staphylococcus cereus flavus**, the cocci of which are arranged in minute clumps (fig. 75). The former produces upon the gelatine plate white dots which spread out superficially and do not liquefy the medium. In thrust-cultures a white deposit is likewise found which gives the surface the appearance of a drop of stearine (fig. 76). On agar a layer spreads out from the

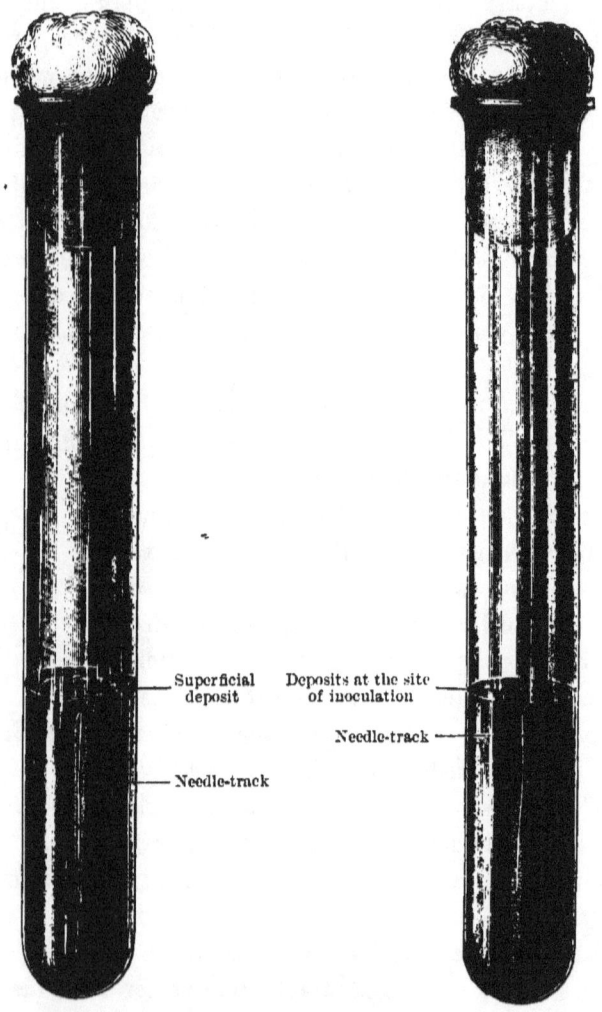

FIG. 76.—THRUST-CULTURE IN GELATINE OF STAPHYLOCOCCUS CEREUS FLAVUS.

FIG. 77.—THRUST-CULTURE IN GELATINE OF STREPTOCOCCUS PYOGENES.

streak, in the neighbourhood of which further colonies are seen. The deposit on potato is similar, but soon becomes dirty grey.

Staphylococcus cereus flavus is differentiated by the yellow colour exhibited in its cultures. In addition to these, Von Schrötter and F. Winkler discovered also the **Staphylococcus cereus aureus**, distinguished from both the others by the orange-red colour of the colonies, and by its slower growth. These microbes are also constantly found in the nasal secretion in cases of coryza. The *Staphylococcus cereus* may also occur in the interior of pus corpuscles; but the cells may then be arranged in diplococci, and in the pus of urethritis it may be confounded with the *gonococcus*. The latter, however, is bleached by Gram's method, whereas the *Staphylococcus cereus* retains its colour.

Streptococcus pyogenes.—In phlegmonous suppurative processes the *Streptococcus pyogenes* described by Rosenbach is constantly present, and is probably identical with the *Streptococcus erysipelatis* of Fehleisen (p. 113). It is found, however, not only in erysipelas but also in puerperal processes occurring in lying-in women. It does not liquefy gelatine, and forms fine punctiform colonies, while in thrust-cultures a thin delicate film appears around the puncture (fig. 77). On agar minute dots resembling dew-drops occur along the streak, and range themselves into a ribbon-like stripe. The elements so far alter upon potato that some individual cells appear larger, and others smaller, than usual. It grows better at incubation temperature than at that of ordinary rooms, but growth is usually slow at best. Transmission of cultures into the lymph-channels sets up a typical erysipelas, according to E. Fraenkel.

The Micrococcus of gonorrhœa.—The infectious nature of gonorrhœal secretion is due, as shown by Neisser, to cocci which have received the name of *Gonococci*, and which can easily be recognised in the secretion owing to their arrangement in pairs. The individual cells are reniform, and may be found in the pus arranged round the nuclei in the interior of the cells, but not in groups (fig. 78), and are distinguished from similar micrococci which occur in the methra by their being invariably decolorised by Gram's method, although otherwise they take the aniline stains very readily.

If it is wished to demonstrate the gonococci in fresh gonorrhœal secretion, the following is the best mode of procedure :—A drop of the pus is placed upon a cover-glass and spread out lightly, so that the glass is covered with a thin cloudy film ; but the common method of laying two cover-glasses one over the other and then drawing them asunder must not be employed here. The cover-glass is now stained with a concentrated alcoholic solution of methyl blue, rinsed in water, dried, and mounted in Canada balsam.

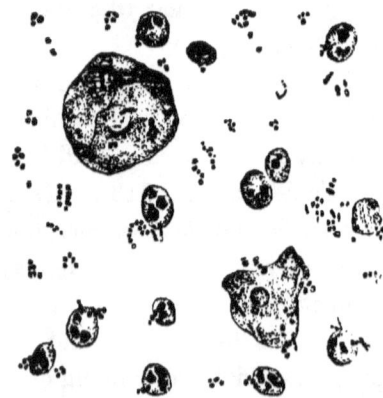

FIG. 78.—GONOCOCCI FROM THE URETHRAL SECRETION. (After Jaksch.)

Neisser has recommended the following method of double staining:—The cover-glass preparations are brought for some minutes into a warm concentrated alcoholic solution of eosine, the superfluous dye is then soaked up with blotting-paper, the preparations laid for a quarter of a minute in concentrated alcoholic solution of methyl blue, and rinsed in water.

Schütz's method consists in preparing a filtered saturated solution of methyl blue in 5 per cent. aqueous carbolic acid, in which the preparations are stained for five to ten minutes in the cold. After rinsing in water they are laid for an instant in very dilute acetic acid, again rinsed in water, and double-stained in very dilute solution of safranine. The gonococci are blue and the pus-cells salmon-coloured.

To differentiate them from other diplococci, Stein-schneider and Galewski recommend that the preparations should be left for half an hour in a solution of gentian violet in aniline water, rinsed, laid for five minutes in solution of potassium iodide, and left in alcohol until decolorised, after which they are again rinsed, dried, and double-stained in dilute alkaline solution of methyl blue. The gonococci are pale, the other diplococci stained blackish.

In making a rapid examination of a series of secretions for the presence of gonococci, F. Winkler uses the following drying-on process:—A clean slide is prepared with a small drop of gonorrhœal secretion, and immediately passed several times through the flame without being previously dried in air. The pus is thus desiccated in a thin film, over which a concentrated alkaline solution of methyl blue is next poured; after a half a minute the preparation is washed in water, dried over the flame, and mounted in Canada balsam. The pus-cells appear pale blue, the gonococci dark blue.

The ordinary nutrient materials which prove favourable to the growth of other micro-organisms fail us completely in the case of gonococcus, which grows neither on gelatine, agar, nor potato, though on the other hand, as Bumm has proved, it thrives on human blood-serum. According to him the cultures exhibit a greyish-yellow layer, resembling from its pointed prominences a group of mountains, and whose margins extend diffusely into the surrounding parts. The serum is not liquefied. The most suitable temperature lies between 33° and 27° C. Growth is very slow, and comes to an end in three days. Cultivation experiments on plover's egg albumen gave good results in the hands of Von Schrötter and F. Winkler, a thin, tolerably transparent, whitish deposit showing itself on the surface of the solidified mass of white of egg in as short a time as six hours at incubation temperature. The growth quickly extends, but ceases after the lapse of only a few days.

Experiments have been carried on by Wertheim at the clinic of Professor F. Schauta in Vienna, on the cultivation of the gonococcus upon a medium consisting of a mixture of two or three parts peptone glycerine agar and one part human blood-serum. He obtains the latter from the blood of the maternal portion of the umbilical cord, which is severed but not ligatured, and the blood caught in sterilised flasks. In this way Wertheim was enabled to grow gonococci in plate-cultures, on which he was able to observe colonies visible to the naked eye even after twenty-four hours in the incubator. The microbes grow rapidly, displaying their characteristic forms and colour, and give the best evidence of their specific action by transmission to human beings. Wertheim's whole process yields astonishing results, and according to his researches growth takes place decidedly quicker when deprived of oxygen than when the gas is admitted.

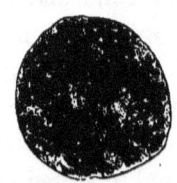

FIG. 79. — DEEP-LYING COLONY OF GONOCOCCUS ON A SERUM AGAR PLATE. (After Wertheim.)

The plate appears diffusely clouded in twenty-four hours, and assumes the appearance of a delicate flocculent layer of moss. The majority of the colonies develop in the substance of the medium, the deep ones appearing whitish-grey by direct, yellowish-brown by transmitted light, and in three days show a peculiar nodulation, which is so regular as to recall the appearance of a blackberry (fig. 79). The superficial colonies have a compact dot situated exactly in the centre, which, when examined with a low power, is found to correspond in structure with the deeper colonies, but is surrounded on all sides by a superficial film which, though at first very delicate, transparent, finely granulated, and colourless, develops in three days round the central mass numerous minute accumulations of condensed matter having a brown colour (fig. 80).

Bacillus of syphilis.—In tissues affected with syphilitic disease and in the layer of pus covering syphilitic ulcers Lustgarten found small comma-bacilli showing a slightly S-shaped curvature, but they have not as yet been cultivated outside the body. In the interior of the tissues they are not found in the interstices of fibrous bundles or other tissue elements, but are included in cells which are much larger than the white blood-corpuscles. The methods of staining are of particular interest. They are carried out in the case of tissues only with fine sections prepared from them; but

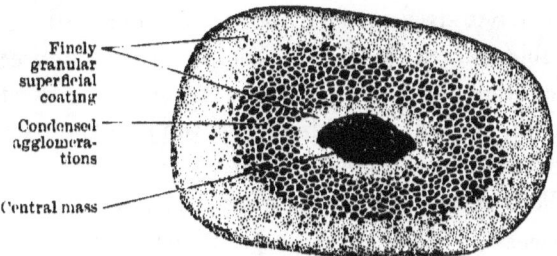

FIG. 80.—SUPERFICIAL COLONY OF GONOCOCCUS ON A SERUM AGAR PLATE.
(After Wertheim.)

in the case of secretions a portion is placed upon a cover-glass and spread out by rubbing. Staining is then done with Ehrlich's solution of gentian violet in aniline water, in which the preparations are allowed to remain for twenty-four hours at ordinary temperature, or for two hours at that of the incubator. They are then placed in absolute alcohol, and after that for ten seconds in a 1½ per cent. solution of potassium permanganate, which deprives the tissues of their colour while the bacilli retain their violet tint. To remove the precipitate of manganese dioxide which forms upon the specimens, they are immersed in an aqueous solution of sulphurous acid. The preparation can be mounted in Canada balsam in the ordinary way, and examined under the microscope. The stained bacilli are decolorised at once by treatment with glacial acetic acid; other acids take longer.

A better method than Lustgarten's is that described by De Giacomi, in which the preparations are treated with solution of iron perchloride after being stained in Ehrlich's solution of fuchsine in aniline water. Lewy recommends staining in carbolic fuchsine and decolorising with distilled water as the surest and most convenient method. That of Doutrelepont and Schütz consists in staining the preparations for twenty-four hours in a one per cent. solution of methyl violet and decolorising for some seconds in dilute nitric acid, after which they are transferred for ten minutes to alcohol of about sixty per cent. strength. The sections are then double-stained by leaving them for a few minutes in a watery solution of safranine, and are thoroughly washed in sixty per cent. alcohol. By this process the bacilli appear blue and the nuclei and tissue light red.

Marschalko recommends staining the sections of syphilitic tissue and cover-glass preparations of syphilitic secretion in Löffler's methyl blue for three to four hours at incubation temperature, or twelve to twenty-four at that of an ordinary room, washing in water, and double-staining for one to five minutes in concentrated aqueous solution of vesuvine.[1]

Bacillus tuberculosis.—*Tuberculosis* was regarded as an infectious disease even by the older physicians; indeed, this opinion may have preceded the teaching introduced in the course of time by the pathological anatomists (such as Virchow and Rokitansky), which was restricted to the description of the microscopic and naked-eye appearances

[1] [Some authorities believe this to be identical with a bacillus found in normal smegma præputiale, &c., which has a similar appearance and similar peculiarities of staining. The latter is said, however, not to stain by Doutrelepont's method.]—TR.

produced by it. But it was Koch who first discovered the cause of the disease in the *tubercle bacilli*, recognisable in all such products of the bodies of men and animals as have undergone tubercular changes. These furnish a foundation for the doctrine that tuberculosis is transmissible, inasmuch as Koch was able to set up a typical tuberculosis experimentally in animals by means of pure cultures. Lortet and Despeignes are of opinion that the tubercle bacilli are frequently disseminated by earth-worms.

Tubercle bacilli are fine rods having a length nearly equal to the diameter of a human blood-corpuscle, somewhat curved, and often united in groups of two or more, but seldom in very extensive combinations. They are destitute of motility. They form spores of an oval outline, both in artificial cultures and in the bodies of animals, and offer a high degree of resistance to drying, boiling, and the action of the gastric juice and of putrefaction. The presence of carbohydrates or of glycerine is, according to Hammerschlag, essential for their growth.

The following is the procedure to be followed when it is wished to obtain a pure culture of tubercle bacillus :—The sputum of phthisical patients is mixed with water, and injected into a subcutaneous sac made over the abdomen in several guinea-pigs. In three or four weeks the tuberculosis will be so far advanced as to cause the death of one or other of the animals, and in the post-mortem examination tubercular changes will be found in the various organs, particularly the omentum, spleen, and liver. Another of the animals is now killed by strangling, a window is cut in the thoracic wall with all due precautions regarding disinfection, and a corner of the lung drawn out with the platinum wire. From this piece of lung some distinct tubercular nodules are now taken and rubbed between two slides. (Of course all the instruments used, such as knives, scissors, and slides, must be carefully sterilised.) The glasses infected with the crushed tubercular masses are now laid upon serum which has been poured out into glass capsules and inspissated, the tubercular mass is smeared over the surface of the serum with a strong platinum wire, and the capsules are covered with glass plates and placed in the incubator. The colonies are perfectly formed in about three weeks, when further cultures can be made from them in test-tubes. When examined under a low power, the cultures show figures coiled in the shape of an S and thickened in the centre, which consist of bacilli massed together (fig. 81). The colonies form dry, white scales, having a dull surface and not larger than poppy-seeds, which adhere but loosely to the surface of the medium, never penetrate into its substance, and do not liquefy it (fig. 82).

FIG. 81.—IMPRESSION PREPARATION FROM A CULTURE OF TUBERCLE BACILLUS ON SERUM.

Pastor gives the following process for obtaining pure cultures of tubercle bacilli from sputum:—A patient is chosen whose sputum is very rich in bacilli and shows comparatively little contamination with other micro-organisms, and he is made to rinse out his mouth and pharyngeal cavity repeatedly with distilled water, and then to expectorate into a sterilised test-glass. The sputum, or more properly the liquid contents of the pulmonary cavities, is shaken up with sterilised water and filtered through fine gauze to remove the coarser particles. A few drops of the filtrate are mixed with melted nutrient gelatine in such a manner as not to render it very turbid, and the mixture is poured out on plates which are left under bell-glasses at room temperature. In from three to four days the various colonies of bacteria contaminating the sputum form. The portions of gelatine between the colonies, and which still remain clear, are now sought out with a lens, carefully cut away with a sterilised knife, and transferred to the surface of serum which has been made to solidify in a slanting position.

The following is Kitasato's method of preparing pure cultures from sputum :— The patient is requested to evacuate his morning sputum into sterilised double capsules. A flake derived from the deeper parts of the respiratory apparatus is isolated with sterilised instruments, and carefully washed in at least ten watch-glasses full of distilled water, one after the other, to remove the bacteria taken up in passing through the cavity of the mouth. The flake is now transferred to glycerine agar or serum. After it has been kept for some fourteen days in the incubator the first colonies form, appearing as circular pure white transparent specks, which project above the surface of the medium. Obtained by this method the colonies are flat, shining, and smooth, whereas those obtained by the earlier method from tubercular organs are from the first dry, dull, and wrinkled.[1] The difference, however, disappears as growth advances, and the whole of the medium becomes covered with a coat consisting of trains of bacilli many times coiled.

Hammerschlag obtains a luxuriant growth by the addition of mannite and

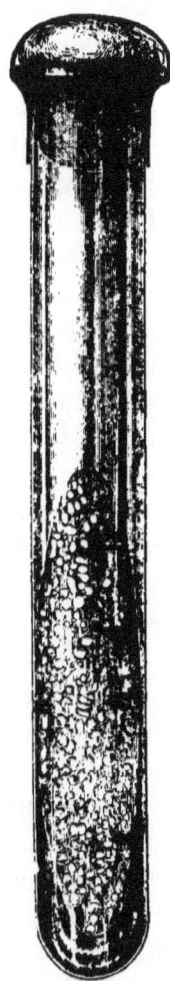

FIG. 82.—CULTURE OF TUBERCLE BACILLUS ON SERUM. (After Baumgarten.)

---

[1] [The bacilli of tuberculosis in birds form a permanent wrinkled membrane on the medium; but they differ so much in other respects also that,

P

grape-sugar to the agar. Pawlowsky found that white detachable colonies appear on the surface of potatoes in twelve to twenty days, but the potato discs must be protected from evaporation in air-tight glass tubes. The bacilli grow in the most luxuriant manner, according to Koch, in an infusion of veal rendered feebly alkaline, and to which an addition of 4 or 5 per cent. glycerine and 1 per cent. peptone has been made. Inoculation is done by floating a fairly large piece of the seed-culture upon the surface of the fluid, and when the cultivation has been kept for several weeks in the incubator, the surface becomes covered with a tolerably thick skin, dry above, and often wrinkled, which a few weeks later becomes moistened by the fluid, breaks up into single ragged pieces, and sinks to the bottom. Growth requires from six to eight weeks for its completion.

With regard to the staining of tubercle bacilli, it is of considerable importance, according to Koch, that the colouring fluid should have an alkaline reaction, as these microorganisms will only take the aniline dyes when acted on simultaneously by alkalies. In order to protect the observer, Pampukes recommends that the tubercular sputa should be sterilised at 120° C. before being examined, as this does not impair their power of absorbing stain.

For practical investigation, the observation of Dahme is of importance, that the flakes which lie at the bottom of the sputum contain the greatest number of bacilli (fig. 83).

According to the method of Koch and Ehrlich, the sputum to be examined is poured upon a dark surface, and the tenacious yellowish particles are picked out. By means of a penholder containing a nib one half of which has been broken away the smallest possible lump of

although identical in appearance and staining, they are now generally regarded as a distinct variety.]—Tr.

sputum is taken up, and is then rubbed between two cover-glasses. When the sputum is very viscid, the cover-glasses between which the lump is spread out are, before being drawn apart, laid on a hot plate at a temperature below 100° C. until a slight clouding takes place, betokening coagulation. The cover-glass preparation, after drying, is passed three times through the flame, and stained in a warm solution of fuchsine or methyl violet in aniline water for a quarter to half an hour. To decolorise, the cover-glass is brought into an acid mixture containing one part nitric acid, two parts water, and two parts sulphanilic acid, or into

FIG. 83. TUBERCLE BACILLI IN SPUTUM. (After Jaksch.)

a solution of 3 per cent. of hydrochloric acid in 9 per cent. alcohol, after which it is rinsed in 60 per cent. alcohol and double-stained in methyl blue or malachite green if the tubercle bacilli have been stained red, or in vesuvine or Bismarck brown if they have been stained blue. It is then rinsed again, dried, and mounted in Canada balsam.

Kaatzer has modified the process in the following manner :—The sputum—best that first expectorated in the morning—having been received into a completely empty spitting-cup, is spread out upon a black plate or a piece of black glazed paper. A very minute particle is now picked up from a yellow purulent spot by means of a platinum needle with the point flattened like that of a scalpel, and which has previously been sterilised at a red heat; this particle is rubbed between two cover-glasses, which are then passed three times through the flame. Staining is done with an aniline water solution of gentian violet which has first been warmed, and the preparations are decolorised by transferring them from the gentian violet, as soon as it has grown cold, to a hydrochloric acid alcohol containing 20 parts water and 2 parts hydrochloric acid to 100 parts alcohol. The preparations remain in this mixture for half a minute to a minute, are then transferred to concentrated alcohol for the same length of time, and afterwards rinsed in water. They are now dried and placed for a half to one minute in a filtered concentrated solution of vesuvine in water by way of double-staining, rinsed in 96 per cent. alcohol, and brought under the microscope in a drop of water. The tubercle bacilli appear of a dark violet colour, or even nearly black, as the aniline brown absorbs the blue part of the spectrum; so that a blue object must seem black in a brown solution. The other portions of the preparation are brown.

The following is the method employed by Günther:— The cover-glass, having been prepared with sputum and fixed in the flame, is deposited face downwards in a watch-glass filled with solution of fuchsine in aniline water, the centre of which is now heated over a *very small* flame while being kept moving vertically up and down. When the fluid begins to give off bubbles, heating is stopped, the watch-glass placed on the table, and let stand for a minute. The process of heating is repeated about five times with intervals of one minute's standing, after which the cover-glass is taken from the stain and laid with the film of sputum uppermost in a watch-glass containing 3 per cent.

hydrochloric acid in alcohol, in which it is moved to and fro for one minute, and is then rinsed in water. Some drops of a dilute aqueous solution of methyl blue are now poured upon it, when, after again washing and drying, it is passed three times through the flame and mounted in Canada balsam.

In the Ziehl-Neelsen process, the cover-glass with the sputum is seized in a forceps, covered with an alcoholic carbolic fuchsine solution, warmed over a feeble spirit flame until bubbles appear, washed in water, and flowed with a 5 per cent. solution of sulphuric acid. It is then rinsed in 70 per cent. alcohol, dried, and double-stained with aqueous solution of methyl blue or malachite green.

According to Friedländer's method, some sputum is smeared on two slides, which are drawn three times through the flame. The sputum is then covered with two or three drops of carbolic fuchsine, passed afresh through the flame, moistened with water and a few drops of nitric acid alcohol (3 per cent. of nitric acid in 90 per cent. alcohol), rinsed in water and watery solution of methyl blue, and examined *without a cover-glass* under the oil-immersion objective.

Kühne recommends shaking up the sputum thoroughly in a glass with a concentrated solution of borax, so as to render it fluid. The cover-glasses prepared with this are stained in carbolic fuchsine for five minutes, decolorised in 30 per cent. nitric acid, rinsed in water, dried, and examined in a drop of aniline oil coloured light yellow with picric acid, which is best done by adding two or three drops of a concentrated solution of picric acid in aniline oil to a watch-glass full of pure aniline oil. To obtain permanent preparations the double-staining should be done by transference for a few minutes to an aqueous solution of picric acid after decolorising with nitric, after which the specimen is dried and put up in balsam. An addition of 4 per cent. citric acid is advisable, in order to increase the solubility of picric acid in water.'

In the process of B. Fränkel, and in that of Gabbet, the staining and decolorisation are carried on simultaneously. The former is done with fuchsine in aniline water, from which the preparation is directly transferred into a filtered mixture of 50 parts alcohol, 30 parts water, 20 parts nitric acid, and as much methyl blue as will dissolve with repeated shaking. If aniline water gentian violet has been used for staining, the preparation is transferred to a vesuvine solution, consisting of a filtered mixture of 70 parts alcohol, 30 parts nitric acid, and as much vesuvine as will dissolve. Staining is finished in a short time—one or two minutes—and the preparation is rinsed in water or alcohol rendered feebly acid with acetic acid, and thoroughly dried.

In Gabbet's modification, the Ziehl-Neelsen carbolic fuchsine is used instead of the aniline water solution, and as the second stain methyl blue in sulphuric acid, consisting of 1 or 2 parts methyl blue to 100 parts of 25 per cent. sulphuric acid.

Arens pours three drops of absolute alcohol over a crystal of fuchsine about the size of a millet-seed in a watch-glass, so as to obtain a saturated alcoholic solution, which is mixed with 2 or 3 c.cm. chloroform. This causes a turbidity of the solution, which begins to clear with separation of the fuchsine in the form of flakes. Cover-glass preparations treated in the usual way are laid in the solution, when clear, for four to six minutes, and, after allowing the chloroform to evaporate, are decolorised in a watch-glass full of 96 per cent. alcohol, to which three drops of hydrochloric acid have been added. They are now rinsed in water and examined in the same medium, or finally double-stained with methyl blue.

Gibbes gave the following recipe : 2 grms. fuchsine and 1 grm. methyl blue are slowly introduced into a solution of 3 c.cm. aniline oil in 15 c.cm. absolute alcohol until they are completely dissolved, and 15 c.cm. water are then added. A few drops of this liquid are heated in a test-tube and poured out into a watch-glass, in which the cover-glass is laid for five minutes, being then washed in alcohol until no more colour is discharged. The bacilli are red on a blue ground. Further staining may be well done with a concentrated aqueous solution of eosine.

Occasionally it is of advantage to subject the cover-glass preparations to different methods of examination. Kaatzer recommends staining the same preparations first with carbolic fuchsine after the Ziehl-Neelsen or B. Fränkel's method, and examining them with the oil immersion lens, after which the cedar oil should be removed with blotting-paper and alcohol, the preparation dried in the flame, stained in hot aniline water gentian violet, decolorised in hydrochloric acid and alcohol, and double-stained in vesuvine. The tubercle bacilli, which were red when first examined, are turned dark violet on a light-brown ground by the second staining.

In cases where staining yields doubtful results, the process of Baumgarten may be employed as a control observation. The preparation of sputum, having been dried in air and passed three times through the flame, is first laid for a few minutes in a watch-glass containing chloroform to free it from fat, rinsed in absolute alcohol, and, after this has evaporated, is placed on a slide in a drop of potash solution, prepared by adding two or three drops of a 33 per cent. solution of caustic potash to a watch-glass of water. If the microscopic examination (without oil immersion) reveals the presence of bacilli, the cover-glass is removed from the slide, dried in the air, passed three times through the flame, moistened with a few drops of dilute aqueous solution of methyl violet, and examined at once under the microscope. The tubercle bacilli, if present in the preparation, remain completely colourless if the examination is not protracted beyond five or ten minutes, whereas all the other bacteria almost instantly assume a blue colour.

In very doubtful cases in which the ordinary methods of staining are at fault, preparations must be made from the sediment. According to Stroschein's method, a tablespoonful of sputum is vigorously shaken in a test-glass with three tablespoonfuls of a mixture of 1 part concentrated solution of boric acid and 3 parts water until the sputum is liquefied, when it is poured into a glass ending in a point below. After standing for twenty-four hours, the clear supernatant liquid is poured off and a cover-glass preparation made from the sediment at the bottom.

Biedert recommends that a tablespoonful of sputum should be boiled with two tablespoonfuls of water and seven or eight of solution of caustic potash; four more tablespoonfuls of water are then added, and boiling is repeated until the mass has become evenly fluid. This remains standing for three or four days in a pointed glass covered over, and is then poured off until a depth of only five to eight millimeters is left. The sediment is energetically stirred up with some fresh white of egg, and some of the mixture is rubbed on a cover-glass, dried at a moderate heat, and stained by the Ziehl-Neelsen method. A small quantity of untreated sputum from the same source may be used instead of the white of egg for fixing the sediment to the cover-glass.

Sedimentation is particularly important in the examination of fluids which are poor in corpuscular elements, and is most conveniently done by centrifuging. In the absence, however, of an apparatus for this purpose, Des Vos mixes white of egg with four times its bulk of distilled water, which causes the globulines to sink to the bottom as a coarsely flocculent mass. Up to 10 c.cm. of the supernatant opalescent liquor, which consists of dilute albumen, are now added to the fluid to be examined, and the whole is well shaken up and heated in the water-bath until the white of egg coagulates. In a short time, according to the quantity of white of egg added, a greater or less amount of a finely flocculent sediment forms, which is to be examined for bacilli.

To examine *milk* for tubercle bacilli, the most convenient mode is to place a drop of the suspected milk upon a cover-glass and add to it two or three drops of a 1 per cent. solution of sodium carbonate. The whole is well mixed with a platinum needle, and the cover-glass is then warmed carefully over a small flame until complete evaporation has taken place. The fat is in this way saponified, so that finally a thin film of soap remains behind on the glass, which is stained, &c., like an ordinary cover-glass preparation.

Regarding the staining of tubercle bacilli in sections of tissue, the same methods hold good with suitable modification, for a more detailed description of which the reader may refer to the general section (see p. 84 *et seq.*). Here it is only necessary to remark that the usual method of staining employed is that with Ziehl's solution, but the sections must be left lying in it for an hour or even longer. Bleaching is carried out in 10 per cent. nitric acid for a half to one minute until the red colour has vanished, after which the sections are transferred to 70 per cent. alcohol until they acquire a rose-red tint or become completely colourless. They are then double-stained in methyl blue, thoroughly washed in alcohol, and put up in Canada balsam. The bacilli are distributed through the tissue, or lie in masses between the lymph cells, or are taken into their interior, and they are also found enclosed in the so-called *giant-cells.*

In order to obtain the active principle *tuberculine*, Koch evaporated pure cultures of tubercle bacilli, made in glycerine and veal infusion, to a tenth part of their volume in a water-bath, and filtered the fluid through an earthenware filter. An albumose can be procured from tuberculine after the process of Klebs, which this observer has named *tuberculocidine*. The tuberculine is treated with platinum chloride, or with the so-called *alkaloid reagents*, and the albumose remaining in the solution thus formed is precipitated out with alcohol.

Bacillus of glanders.—This bacillus was discovered by Löffler and Schütz to. be the cause of the disease called *glanders* or *malleus*, which may be transmitted from horses and donkeys to human beings. The micro-organisms are also found in what are called *glanders nodules*, from which they penetrate into the surrounding parts and originate morbid phenomena, and which occur as a rule in the cavity of the nose, and form deep ulcers on the mucous membrane covering the turbinated bones, on that of the larynx, and in the lungs. The neighbouring lymphatic glands are swollen and rendered hard, as well as the cutaneous lymphatic vessels, which often burst outwards, forming ulcers.

The bacilli are slender rods with rounded ends. They are about the size of tubercle bacilli, are without power of automatic movement, and grow upon our nutrient media only at incubation temperature. Development is suspended at a temperature above 42° C. In consequence of this peculiarity gelatine cannot well be used, as growth is very scanty at a lower temperature, but the micro-organisms thrive very well on glycerine agar at 37° C., showing pale

BACILLUS OF GLANDERS

yellow or whitish round colonies on the plate as early as the second day, while along a streak of inoculation made on the same medium a shining coating appears in four to five days. On the surface of blood-serum there are developed yellowish deposits which coalesce with one another and form a slimy covering. The cultivation on potato is particularly important: a deposit appears at incubation temperature which is at first yellowish and transparent, but which gradually increases and assumes a darker tint. In about a week the colour changes to reddish-brown, and finally passes into red. The bacilli readily take up the aniline dyes, but discharge them again when Gram's method is employed (fig. 84).

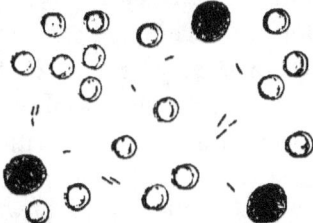

FIG. 84.—BACILLI OF GLANDERS IN HUMAN BLOOD. (After Jaksch.)

The staining process of Löffler consists in colouring the cover-glass preparations for five minutes in a hot alkaline solution of gentian violet or fuchsine in aniline water containing an addition of 1 per cent. of caustic soda solution. Bleaching is done in 1 per cent. acetic acid coloured to a wine yellow with a watery solution of tropæoline OO, in which the preparations remain for an hour, and are then rinsed in water.

Kühne treats the cover-glasses with hot carbolic fuchsine or carbolic methyl blue, and decolorises in water containing two drops of hydrochloric acid per 100 grams.

For staining glanders bacilli in sections, Kühne has devised the following method, which is reliable and easy in use:—The sections, having been stained with carbolic fuchsine and soaked in water, are quickly decolorised in an aqueous solution of hydrochloric acid (3 per cent.), well rinsed in water, and then either dipped lightly into alcohol or freed as far as possible from water by pressing them with blotting-paper Aniline oil mixed with 20 per cent. of oil of turpentine is now allowed to act on the sections in a watch-glass for eight to ten minutes, during which they remain adherent to the cover-glasses, and thence they are transferred to oil of turpentine, xylol, and balsam.

Noniewicz recommends that the sections should be stained for a few minutes in alkaline methyl blue, rinsed in water, and decolorised in a mixture of 3 parts of $\frac{1}{2}$ per cent. acetic acid, with 1 part of a $\frac{1}{2}$ per cent. solution of tropæoline OO in water. Thin sections are only dipped in quickly, but thick ones may remain in for two to five seconds or longer. They are then washed in water, spread out upon a slide, freed from moisture with blotting-paper, dried in air, cleared with xylol, and mounted in Canada balsam. The glanders bacilli appear black on a more or less blue ground. By this method the discovery was made that the characteristic bacilli appear in the acute form of the disease, whereas in the chronic form they are sparser, and round bodies, which stain intensely, appear in their stead in large numbers.

The glanders bacilli are also found in the interior of dead cells. The vessels are free, and but few appear in the blood. Transmission experiments made with pure cultures have yielded successful results, the infection spreading most from wounds in the skin. The communicability of the disease is considerable, and the infecting power of the virus is not altered by drying for three months.

Other microbes of pus.—It need hardly be said that the pus from wounds in other specific infectious processes contains the corresponding micro-organisms, e.g. *Bacillus anthracis* and the bacillus of tetanus; but besides these, a considerable number of other microbes have been found which have the power of originating suppurative processes. Thus Weichselbaum discovered the **Micrococcus intracellularis meningitidis**; Neumann and Schäffer the **Micrococcus meningitidis purulentæ**; Becker the **Micrococcus osteomyelitidis**, and Heydenreich the micrococcus of Pende's ulcer (**Micrococcus Biskra**); while the **Streptococcus contagiosæ coryzæ equorum** was detected by Schütz in suppuration occurring in animals, and Pfeiffer found in the abdominal cavity of a guinea-pig which died spontaneously a tenacious, puriform exudation consisting of a pure culture of **Bacillus capsulatus**.

Some of the methods invented by Unna for staining micro-organisms in cutaneous abscesses (see p. 90) are excellently adapted for showing those in pus. The following process may be especially recommended:—The cover-glass preparations are stained in carmine, transferred for two minutes to borax methyl blue (1 part each borax and methyl blue to 100 of water), rinsed in water, and placed in alcohol to which a few drops of Spiritus saponis kalini [1] have been added. They are finally brought once more into alcohol, and may then be submitted to examination.

[1] [See p. 91, note.]- Tr.

CHAPTER X

BACTERIOLOGICAL EXAMINATION OF THE ORGANS AND CAVITIES OF THE BODY AND THEIR CONTENTS

Micro-organisms of the living body.—In this section a series of micro-organisms will be dealt with, which find a congenial pabulum in the organs and cavities of the living individual. They are derived partly from the air, partly from water or other surrounding media, and may occur either as saprophytes or parasites.

I. THE SKIN

Micro-organisms of the skin.—Most of the micro-organisms met with on the skin in the cutaneous secretions are saprophytes, but many pathogenic varieties also occur in this situation. The space beneath the nails, in particular, is rich in forms, containing especially germs, according to Preindlsberger, which are found also on other regions of the cutaneous surface as well as often in the air, and hence may penetrate with it into the respiratory organs. Pathogenic germs are found amongst them. Penetration of micro-organisms into the unbroken skin probably takes place by the openings of the glands, or perhaps by the agency of stinging insects.

Amongst the cutaneous micro-organisms the bacteria of the female generative organs are also included. Straus and Toledo have found that microbes do not occur either in the parietes or secretion of the normal uterus, and it is only in the vagina, where there are accumulations of epithelial masses, that the bacteria find a favourable medium upon which to develop. In normal vaginal secretion the *vaginal bacillus* of Döderlein is regularly found, and the *thrush fungus* and *yeast cells* are often met with, while *Staphylococcus pyogenes albus*, *citreus*, and *aureus* are no rarities.

Micro-organisms are also found in the sebaceous and sweat glands. Brunner discovered, for example, that streptococci may be detected in the sweat during suppurative processes, and Garré conversely was able to start a phlegmon by rubbing a culture of staphylococci into his unwounded arm. The appearance of *coloured sweat* is due to the presence of *Bacillus pyocyaneus* and *Micrococcus hæmatodes*, whilst in the foul-smelling perspiration on the feet Rosenbach found the *Bacillus saprogenes II.* (see p. 176).

Preindlsberger isolated from the sub-ungual space the *Micrococcus cereus albus* and *flavus*, the *Diplococcus liquefaciens albus* and *citreus*, *Micrococcus candicans*, *Sarcina alba* and *flava*, and *white yeast*; and of pathogenic micro-organisms the *Staphylococcus pyogenes aureus* and *Streptococcus pyogenes*.[1]

Methods of examination.—In order to stain the micro-organisms in the skin, the objects to be examined must first be freed from fat in alcohol and ether, after which they are stained in glycerine methyl blue by spreading them out with needles upon a slide and letting the stain act for some minutes ; the micro-organisms appear blue, the epidermis colourless. In Böck's process the epidermic scales, after being freed from fat, are deposited for half a minute to several minutes in borax methyl blue, then for half a minute to a minute in a weak aqueous solution ($\frac{1}{2}$ to 1 per cent.) of resorcine, and then for some minutes to an hour in alcohol. To make the fungi stand out clearly, the

---

[1] [The *Trichophyton tonsurans* also occurs in this situation, causing the disease known as *onychomycosis*.]—Tr.

epidermis is decolorised for some seconds in a weak solution of hydrogen peroxide and then transferred to alcohol, xylol, and Canada balsam.

Unna devised the following method for the rapid recognition of fungi in epidermic scales:—The horny scale to be examined (crust or comedo) is laid upon a slide, moistened with a drop of acetic acid, and rubbed to a pulp by means of another slide placed crosswise upon the first, after which the two slides are separated and dried rapidly over the flame, so that all the acid is evaporated off again. Some drops of ether and alcohol are now poured upon the free upper end of the slide, in order to wash out the fat, and then two drops of solution of borax and methyl blue are at once placed upon one slide, which is again covered crosswise with the other, and the pair are held over the flame for ten to twenty seconds. The preparations are now either immediately subjected to a further process of bleaching, or are dried over the flame. Styrone, glycol, or equal parts of glycerine and ether are used as bleaching agents. By the first or styrone method the preparations stained in methyl blue are rinsed in alcohol, decolorised for two minutes in styrone, again rinsed in xylol, and put up in balsam; by the second method, the stained preparations are rinsed in water, decolorised in glycol for two to five minutes, rinsed again in water and then in alcohol, dried over the flame, and mounted in balsam. One per cent. solution of acetic, oxalic, citric, or arsenic acid, or of hydroxylamine, or one per cent. aqueous solution of soap, one per cent. salt solution, five per cent. aqueous solution of resorcine, and one per cent. alcoholic solution of hydrochinone, may be also used for decolorising.

Diplococcus subflavus.—Bumm very frequently found in the vaginal secretion diplococci closely resembling the gonococcus in appearance. They thrive fairly well upon our culture media, but grow slowly; and gelatine is liquefied, also slowly. Not until a week has elapsed do sulphur-yellow punctiform colonies, whose border appears fibrous under low powers, show themselves on the gelatine plate. In thrust-cultures a yellowish raised deposit with a dull surface appears after three days, and gradually spreads; no growth can be seen along the needle-track. A greyish-white raised layer develops upon agar even in twenty-four hours, and assumes a yellow colour after some days. On potato a smooth, spherical, sharply-defined deposit forms, which adheres to the needle and is very viscid. Serum is liquefied.

Irrespective of these biological peculiarities, the *Diplococcus subflavus* may be distinguished from the *gonococcus* by the fact that it does not give up its stain under Gram's process.

Micrococcus lacteus faviformis.—Bumm has very often found in the vaginal secretion diplococci which show a tremulous motion. When examined in the hanging drop they exhibit a propensity to unite in a clump which forms the figure of a honeycomb. Gelatine is not liquefied, and the colonies on the plate are small and evenly circular. Little white dots form upon thrust-cultures even in one day, and later unite into milk-white islets. A milk-white layer develops also upon agar and potato. The diplococci do not surrender their colour under the application of Gram's process.

Diplococcus albicans amplus.—A not uncommon guest in the vaginal secretion is the coccus of this name discovered by Bumm, the elements of which are differentiated from the rest of the diplococci by their considerable size. Gelatine is slowly liquefied, and the gelatine plate shows greyish-white, somewhat elevated islets. In thrust-cultures a moist coating appears on the surface and a white stripe along the needle-track. On agar, a stripe-like grey deposit develops rapidly at 35° C.

Diplococcus albicans tardus.—Unna and Tommasoli found this microbe in cases of *eczema*. The diplococci are immotile, sometimes arranged in clumps and chains, and do not liquefy gelatine. Roundish colonies form on plate-cultures, presenting an uneven lumpy surface when examined under the microscope. The colonies gradually become greyish-yellow, lighter in the centre and at the margin, so that they appear girdled by a light zone. Thrust-cultures show a superficial yellow coating, surface-cultures on agar a greyish-yellow stripe with uneven edges.

Diplococcus citreus liquefaciens.—In cases of eczema Unna and Tommasoli also found immotile cocci which liquefy gelatine by their growth. Whitish dots develop on the plate, which rapidly become fluid and assume a lemon-yellow colour. Thrust-cultures show a yellowish coating at first, but after a few weeks a portion of the lemon-yellow culture is seen floating in the liquefied mass, while the rest, which has also a yellow tint, is found at the bottom. The lemon-yellow colour on the surface of the fluid mass is particularly characteristic. A yellowish layer forms upon agar, and also on potato.

Diplococcus flavus liquefaciens tardus.—Besides the micro-organisms just described, the same observers found in cases of eczema diplococci shaped like a roll of bread, which liquefy the gelatine rather slowly. Punctiform brownish-yellow colonies form on the plates, and yellowish masses float on the surface of the gelatine when it has become fluid. In thrust-cultures also liquefaction progresses, but with such extreme slowness that two months elapse before half a cubic centimeter of gelatine is liquefied. A thick slimy coat of greenish-yellow colour develops on agar, and a lumpy sulphur-yellow deposit upon potato.

Micrococcus hæmatodes.—Babès isolated from the foul-smelling sweat of the axilla, which leaves a red spot upon the under-clothing, a micrococcus the elements of which are circular or oval, and united to one another by means of a transparent jelly-like red mass. Babès cultivated them at incubation temperature upon coagulated white of hens' eggs, on which medium they form blood-red confluent islets.

Micrococcus of trachoma.—Sattler and Michel believe that the cause of the disease of the conjunctiva and cornea known as *trachoma*, or Egyptian ophthalmia, is a diplococcus distinguished from the gonococcus by its smallness and the feeble marking of its line of division. It does not lose its colour under Gram's process, and does not liquefy gelatine. On the plate little white nebulosities develop, and in thrust-cultures a row of globules is seen along the track, resembling a string of pearls, while a shining white deposit, which later becomes yellow, forms on the surface. A grey coating develops upon agar, and a white ribbon-like stripe upon blood-serum. Experimental transmission to the human conjunctiva produces a typical trachoma.

Diplococcus of acute pemphigus.—Demme found in the vesicles of *pemphigus* cocci arranged in pairs, which lie together in dense masses. The micro-organism seems to be pathogenic, since, when pure cultures are transmitted to guinea-pigs, inflammatory nuclei form in the lungs, with emaciation; and injection into the blood produces similar phenomena. The diplococcus only grows at incubation temperature, and is best cultivated upon agar. Round milk-white colonies form on the plate, sending out lateral processes towards the surface and in a radial direction, which cause the islets to show irregular prominences. The appearance of superficial cultures on agar is similar. On serum and potatoes growth is very slow.

Vaginal bacillus.—Döderlein detected slender rods, not endowed with motility, in the normal vaginal secretion of the adult; that of new-born infants is destitute of bacilli. *Vaginal bacilli* are facultative anaerobes, growing better in the absence of oxygen. Since their growth does not advance except at the heat of the body they cannot be cultivated to advantage upon gelatine, but upon agar an exceedingly delicate film develops, consisting of fine minute drops which are clear like water. The culture is very sensitive to dryness and to degrees of temperature lower than that of the body. In the incubator the layer develops into somewhat flat elevations, which always remain clear and transparent. Cultivation succeeds also in bouillon, in milk, and upon blood-serum, but no growth takes place on potato. The bacilli are active acid-formers, and the acid contained in the vagina (0·5 per cent.) is due to their vital energy, the secretion of new-born infants, which is free from them, showing a neutral reaction.

Bacillus of symptomatic anthrax (Rauschbrand).—*Symptomatic anthrax* is a disease occurring in calves and lambs, which are killed by it in two days. The cause has been found by Bollinger, Thomas, Kitasato, Kitt, &c., to be bacilli met with in the sero-sanguineous fluid from the subcutaneous cellular tissue, which resemble cocci or have a drumstick form, possess the power of automatic movement, and belong to the anaerobes. The motility is due to the presence of numerous flagella. The bacilli are decolorised by Gram's process; their growth liquefies gelatine. Warty irregular islets appear on plate-cultures, and in high cultivations development takes place along the needle-track, with generation of gas. The islets appear as globular cavities filled with fluid, from which whitish fibres penetrate into the gelatine. On agar an active generation of gas takes place at incubating heat, and in blood-serum peculiar spirally twisted tress-like figures form, which show the greatest difference in size, and consist, according to Löffler, of flagella which have been torn off and compacted together (fig. 85). The surface of potatoes acquires a moist, polished appearance, in consequence of a thick white coating which can be easily detached. Bouillon cultures smell of rancid butter, as also those on agar.

Transmission to sheep, cattle, and goats is attended with success, but pigeons and mice are nearly immune. Owing to the anaerobic character of the micro-organism, inoculation can only be done subcutaneously. Cultures on solid media retain their virulence longer than those in fluids.

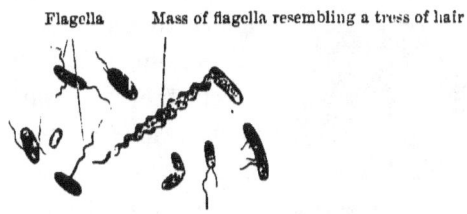

FIG. 85.—BACILLUS OF SYMPTOMATIC ANTHRAX, FROM A CULTURE ON SERUM. Magnified 1,100 times. (After Löffler.)

Lepra bacillus.—Hansen found in leprous tissue, and particularly in the *lepra cells*, small slender rods which have club-shaped ends and approximate nearly to *tubercle bacilli* in size. They are immotile, and are principally distinguished from the tubercle bacilli by their property of staining readily in aqueous solutions of the aniline dyes. Besides this, however, they also stain by the methods devised for tubercle bacilli, as well as by Gram's process, but staining succeeds with much greater ease with bacilli of lepra than with those of tubercle.

By Baumgarten's process, cover-glass preparations are stained for six or seven minutes in a dilute alcoholic solution of fuchsine, decolorised in a mixture of one part of nitric acid and ten of alcohol, rinsed in water, and double-stained in an aqueous solution of methyl blue.

According to Lustgarten, cover-glass preparations, after staining in aniline water solution of fuchsine or gentian violet, are decolorised in one per cent. sodium hypochlorite and rinsed in water.

Unna stains sections for twelve to twenty-four hours in aniline-water fuchsine, and decolorises in a ten to twenty per cent. aqueous solution of nitric acid until the preparations have acquired a yellow colour, after which they are transferred to dilute alcohol, and left there until the tint has turned red. The acid still present is now removed by immersion in a weak solution of ammonia in water, and the water is absorbed up with blotting-paper, after which the preparations are mounted in balsam free from oil, dissolved in chloroform.

FIG. 86. INLET OF LEPRA BACILLUS ON A GELATINE PLATE. (After Macé.)

The Lutz-Unna method consists in staining the preparations in a warm aniline-water solution of gentian violet until they have become darkly-coloured, after which they are decolorised by successive immersion in solution of iodine and potassium iodide, in alcohol and nitric acid, and in plain alcohol. This procedure must be repeated until the preparations show a bluish slate-colour, after which they are mounted in balsam dissolved in oil of thyme or oil of cloves.

Cultivation of the bacilli succeeded in the hands of Bordoni-Uffreduzzi upon peptone glycerine serum, which was inoculated with marrow from the bones of a person who had died of leprosy. From this it was possible to transfer the microbes to gelatine, on which they form rounded colonies at 20° to 25° C. without liquefying the medium (fig. 86). A surface-culture on agar likewise shows roundish colonies, prominent in the centre. Those on serum are ribbon-like with jagged edges, and do not cause liquefaction. Cultivation experiments are best carried on at 37° C.

The disease can be transmitted to men and animals. In the tissues, the bacilli are found in the cutaneous connective tissue, nerve-sheaths, spleen, and lymphatic glands, but do not occur in the blood. According to Unna they are found in the lymph-paths of the tissue, and he also believes the *lepra cells* to be transverse sections of lymphatic vessels.

The other methods of staining the micro-organisms of the epidermis, which can also be used for abscesses in the skin, have already been dealt with in the general section, to which the reader is referred (p. 90).

Bacillus sycosiferus fœtidus was found by Tommasoli on the hairs of the beard of persons suffering from *sycosis*. The rods are short, rounded at the ends, and immotile, and do not liquefy gelatine. Growth makes slow progress at room temperature. On plate-cultures white punctate colonies form, and gradually spread out like a slimy veil. A white nail-head growth appears in thrust-cultures, and a greyish-white veil-like layer in surface-cultures on agar, this being formed by the coalescence of single isolated specks, and often showing wavy stripes. The colonies on potato are raised, and coalesce with the formation of a yellowish-white colour. Pure cultures may be obtained from plucked-out hairs or from the pus in a bacillary sycosis; and if, on the other hand, some of a pure cultivation be rubbed on the skin, there result reddening and intense itch, and vesicles form in the neighbourhood of the hair.

Ascobacillus citreus.—This was found by Unna and Tommasoli in cases of *eczema* of the skin, and consists of motile rods whose propagation but slowly liquefies the gelatine. Dark colonies form on the plate, and in thrust-cultures there appears a lemon-yellow coating, which after some weeks floats on the surface of the liquefied gelatine, while a few greenish flakes show themselves at the bottom. On agar a yellowish layer develops, formed by the confluence of several globular colonies. The growth on potato is more rapid, and in about two weeks the yellow coating shows in the centre a greenish-yellow colour with fine light venules, so that the whole resembles a vine leaf.

Bacillus xerosis.—E. Fränkel, Leber, Neisser, Colomiatti, and several others found in *xerosis of the conjunctiva* short bacilli, devoid of motility, which will not grow on gelatine or potato. A veil-like film develops on agar at incubation temperature, and on serum there forms a focus of rosette-like figure. The bacilli are found in the white fatty-looking scales on the conjunctiva, which consist of horny and fatty degenerated epithelial cells, free fat drops, and the short bacilli.

Trichophyton tonsurans.—Malmsten and Gruby found in the scaly accumulations of *Herpes (tinea) tonsurans* a fungus designated as *Trichophyton tonsurans*, [which occurs in the hair-follicles and hair-substance, and in the epidermis]. Leslie Roberts cultivated it upon infusion of malt containing sugar and on alkaline veal-bouillon. The portion of skin affected by the disease was cleansed with corrosive sublimate, several short stumps of hairs were pulled out with sterilised forceps, and their bulbous extremities snipped off with a pair of scissors (previously sterilised at a glowing heat) in such a way that they fell directly into a corresponding number of flasks prepared with infusion of malt or bouillon. Even in twenty-four hours colourless threads radiate on all sides from the yellow hair-bulb, and grow together into a colourless tuft of mycelium. If they rise to the surface, the part exposed to the action of the air becomes very speedily covered with a snow-white powder, which lends the surface, seen from above, the appearance of a membrane, and which gradually becomes yellow. Already in the earliest stages of development there occur in the mycelial tubes dilatations resembling ampullæ and filled with a granular mass, and these are taken to be the spores of the mycelium. The tubes enlarge and become segmented, which causes a distinct resemblance to a string of pearls; the mycelium now gradually becomes finer, the divisions are small, and show smooth walls with bulgings. Verujski has shown experi-

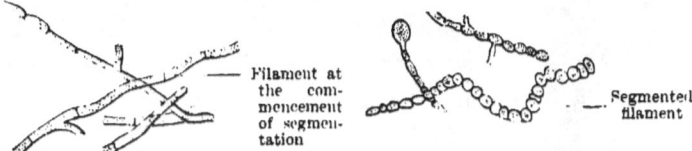

FIG. 87.—MYCELIUM FILAMENTS OF TRICHOPHYTON TONSURANS, FROM A PURE CULTURE ON INFUSION OF MALT. (After Leslie Roberts.)

mentally that glucose, and not, as Grawitz supposed, a nitrogenous body, is the medium suitable for the trichophyton. The development, therefore, of the tonsurans fungus in the epithelium of the skin covering the body does not progress by means of spores, but merely by the swelling, constriction, and final separation of fibres (fig. 87).

This fungus is also, according to H. von Hebra, the exciting cause of *impetigo contagiosa* (an exanthem characterised by the formation of pustules), as well as of *eczema marginatum*, [and it is now agreed that *tinea sycosis*, *tinea circinata* (common ringworm), and *onychomycosis* (an affection of the nails) are due to it. (See also note at end of chapter.)]

Gelatine is quickly and actively liquefied, according to

Grawitz, and the fungus fur floats upon the surface as a thick coating, white above and yellow underneath.

The fungus of favus (Achorion Schœnleinii).—In the scaly accumulations from persons suffering from *favus*, Schönlein discovered a fungus to which the name of *Achorion Schœnleinii* has been given. The favus crusts or *scutula* show the mycelium of the organism, which is remarkable for the fact that the fibres grow up perpendicularly from the horny layer. The right angle formed by these fibres with the substratum is characteristic of the scutula of favus, since, in the horny scales permeated by the *fungus of pityriasis versicolor* or by the *Trichophyton* the fibres of the organism form acute angles, or run parallel with the cells of the horny layer.

Unna differentiates three varieties of favus fungus—the **Achorion euthythrix**, with straight fibres and abundant formation of spores; the **Achorion dicroon**, with forking hyphæ; and the **Achorion atacton**, the hyphæ of which run an irregular course and possess peculiar knotty ramifications and many sharp angles. The first forms voluminous greyish-yellow scutula (*favus griseus*), the second sulphur-yellow scutula, which are slow in developing (*favus sulphureus tardus*), and the third sulphur-yellow scutula also, which form more quickly (*favus sulphureus celereus*). Unna used for cultivation a medium composed of 4 per cent. of agar, 1 per cent. of peptone, and 5 per cent.

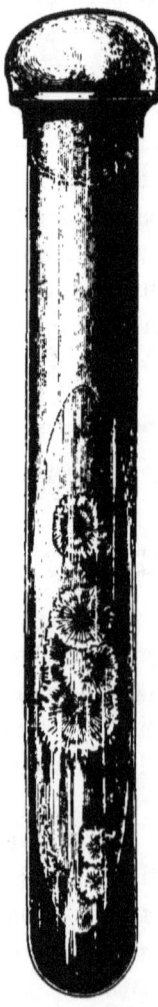

FIG. 88.— SURFACE CULTURE OF THE FUNGUS OF FAVUS ON SERUM FROM OX-BLOOD, NINE DAYS OLD. (After Plaut.)

of levulose or ½ per cent. of common salt. In order to obtain pure cultures dry media only must be employed, and they must accordingly have been kept in a warm place for a considerable time before use. All other micro-organisms which might spoil the pure culture of favus require a tolerably large amount of water, and soon cease to grow upon very dry soils. The *Achorion euthythrix* forms dense woolly white furs, half a centimeter in height, upon the nutrient medium, and grows but a short way into its substance, whereas the growth of *Achorion dicroon* lies for the most part in the medium and only raises a thin coating of mycelium above the surface. The *Achorion atacton* occupies an intermediate position between the other two. Those varieties of favus fungus which are marked by their deep growth in artificial cultures grow also into the hair-follicles and are more difficult to treat.

Gelatine is liquefied very late and only very slightly by the growth of the favus fungi. On agar plates there develop in twenty-four hours at incubation heat, according to Plaut, very small shining fibres resembling cotton wool, which are recognisable under the microscope as minute masses of conidia. In forty-eight hours the colonies have changed to round milk-white opaque discs possessing a border of very fine and quite short threads. On potato Plaut obtained, at incubation temperature, an irregular discoloration on the site of inoculation, from which the fungus grows into the substance of the potato. Similar coatings to that upon potato appear on boiled white of egg, and on blood-serum a greyish-white, thin, coherent film develops below the surface and sends out long thin processes downwards and parallel to it. The serum is neither liquefied nor discoloured (fig. 88).

[**Microsporon furfur** is found in the yellowish or brownish scaly patches on the skin in *pityriasis versicolor*. When some of the scales are examined under the microscope in a drop of potash solution, abundant mycelial threads are seen, together with spores, which are gathered in groups resembling clusters of grapes. Cultivation on solid media has not yet succeeded.]—NOTE BY TRANSLATOR.

Certain observers, most recently M. Sabouraud, believe that *Trichophyton tonsurans* really includes several distinct varieties. The latter states as the result of an extended research that there are two principal kinds, distinguished by the size and arrangement of their spores and by their biological characters. The first, called by him **Trichophyton microsporon**, has spores 3μ in diameter which lie in no particular order, and is apparently destitute of mycelium. It occurs only in *tinea tonsurans*, causing about 60 per cent. of the cases, and is most difficult to treat. On beer-wort and agar it produces a growth which is at first white and downy, but from the fifteenth to the eighteenth day becomes dry, floury, and yellowish ; on potato it is dry and yellowish-brown from the first.

The other variety, **Trichophyton macro-** or **megalosporon**, which causes all the remaining forms of tinea, is distinguished by spores 7μ to 8μ in diameter and arranged in lines in the branches of its mycelium. Cultures show a white downy growth, which appears a little later than in the case of the former variety, and does not alter. On potato a reddish-brown spot, like dried blood, appears ten days before the white down develops. The *macrosporon* variety, however, seems to include many species distinguished by further differences in growth and producing distinct clinical phenomena. One such causes about half the cases of *tinea circinata*, the others are rare.

The above-named varieties always reproduced the same species, and the former could not be made to grow on any part of the body except on the head.

(See *Ann. de Derm. et de Syph.*, vol. iii. Nov. 1892 ; also *Brit. Journ. of Derm.*, vol. v. Jan. 1893, and *Med. Week*, vol. i. Feb. 24, 1893.)—TR.

CHAPTER XI

THE ORGANS AND CAVITIES OF THE BODY AND THEIR CONTENTS (*continued*)

II. THE DIGESTIVE TRACT

The Cavity of the Mouth

Micro-organisms of the mouth and their examination.—A number of micro-organisms are found in the mouth, which are derived from the air and are taken into the digestive tract with the food. Their occurrence in the buccal cavity is so constant that in each examination of saliva cocci, bacilli, spirilla, and other fungi are encountered in large numbers. When saliva is freely smeared over the surface of a cover-glass, allowed to dry, fixed in the flame, and stained in a dilute alcoholic solution of an aniline dye, micro-organisms will be found which belong in part to the *mouth fungi*, the properties of which we know by cultivation, but a large number have not as yet been successfully grown on our nutrient substances. Besides the saliva, the particles of food which remain behind in the mouth, and especially between the teeth, play an important part in the retention of bacteria in this cavity, and to these must also be added changes in the teeth and gums which promote the development of micro-organisms. The importance of the saliva to the life of the micro-organisms lies in the alkaline reaction which it commonly possesses, since if an acid reaction sets in, its capability of serving as a favourable nutrient medium for microbes becomes diminished; but in this case, on the other hand, the teeth become so altered by the acidity as to furnish a suitable nidus for their growth (Miller).

If solution of iodine and potassium iodide be added to a sample of fresh saliva, many fibres are seen which are coloured yellow and intertwined with one another. These are described as **Leptothrix buccalis**. Others stain bluish-violet with tincture of iodine, appear in tufts, and are called the **Bacillus buccalis maximus** (fig. 89). Micrococci which appear bluish-violet after the addition of tincture of iodine bear the name of **Iodococcus**, amongst which Miller distinguishes by cultivation an **Iodococcus magnus** and an **Iodococcus parvus**. Sometimes cocci are stained blue and their sheath yellowish by tincture of iodine (**Iodococcus vaginatus**), and some stain rose-red. Infections of the teeth leading to caries are supposed by Miller to be due to the fact that in acid solutions the enamel loses the lime it contains and grows soft, this softened substance then forming the gelatinous ground-substance which is altered just like gelatine by the action of micro-organisms, so that damage to the tooth results, leading to caries.

FIG. 89.—LEPTOTHRIX BUCCALIS. (After Jaksch.)

Micrococcus salivarius septicus and Bacillus salivarius septicus.—Biondi found both micro-organisms in saliva, the bacillus in healthy as well as diseased individuals, the micrococcus in *puerperal septicæmia*.

The *Micrococcus salivarius septicus* consists of round or sometimes oval cocci, the growth of which does not liquefy gelatine. Upon plates round colonies form which are dark-coloured; in thrust-cultures the growth is granular, owing to the close apposition of the punctiform colonies, and on agar there is an abundant superficial growth.

The *Bacillus salivarius septicus* shows short rods with pointed ends which at times lie in chains, at others in clumps, and grow best when a little phosphoric or hydrochloric acid (0·04 per cent.) is added to the medium. Gelatine is not liquefied. Round colonies with a transparent periphery develop on the plate, and the islets show a zigzag network under a high power. A thrust-culture appears as a clear band with peripheral dots. If a transmission be made to agar from a gelatine culture, a transparent coating develops.

Both have a pathogenic action upon mice, guinea-pigs, and rabbits, and the animals die of *septicæmia*, sometimes in as short a time as forty hours.

Bacillus ulna.—Vignal found in the mucus from his mouth rods with rounded ends, which liquefy gelatine in growing. The colonies are round, and appear on the plate like two concentrically-arranged stripes separated by a clear line. Several zones appear in the liquefied part and towards the border, the outer one of which looks like a tangle of fine fibres. In thrust-cultures a funnel-shaped area of liquefaction appears along the needle-track, on the bottom of which lie crumbling accumulations of bacteria, and in a few days a shining membrane forms on the surface. Growth progresses best at 20° C. On agar there appears at incubation temperature a deposit in the form of a coherent pellicle, which can with difficulty be raised from the surface of the medium. A coating develops on potato at the same temperature. Serum is liquefied. The smell of the cultures resembles that of decomposed pus.

Bacillus gingivæ.—Miller isolated thick rods with rounded ends from the deposit in a mouth not kept properly clean, and in the suppurating pulp of a tooth. They form at ordinary temperature round colonies on the gelatine plate, which liquefy in two days. Thrust-cultures show a funnel-shaped excavation. On agar there forms a greenish-white layer. Intraperitoneal injections into white mice prove fatal with the symptoms of peritonitis, and subcutaneous injections lead to the formation of abscesses.

Bacillus diphtheriæ.—Klebs was the first to draw attention to the presence of bacilli in diphtheritic disease, and Löffler has succeeded in producing similar morbid conditions in animals by the injection of pure cultivations. The bacillus is an immotile straight or slightly curved rod of the same length as the tubercle bacillus, but much thicker. According to Löffler, it grows upon a nutrient medium composed of 3 parts serum of sheep's blood, 1 part neutralised veal bouillon, 1 per cent. of peptone, 1 per cent. of grape sugar, and ½ per cent. of common salt; and Kolisko and Paltauf state that luxuriant growth takes place upon a nutrient bouillon containing sugar, while Kitasato obtained an abundant development upon glycerine agar. It should be observed that the best temperature for all these experiments is 33° to 37° C., and, speaking generally, the diphtheria bacilli require for their development a heat of over 20° C. They retain their vitality even when completely dried, and Löffler found them still capable of developing after 101 days. If diphtheritic membranes are protected from the action of light and kept dry, cultures retaining their virulence can be prepared from them after as long a time as three months. They form upon gelatine small, round, white colonies, with a coarsely granular texture and irregular edges, and do not liquefy the medium. Little white dots may be observed in the thrust-culture, and the rodlets often assume altered shapes. The growth upon ordinary agar is scanty, although luxuriant upon glycerine agar. A greyish-white deposit is visible in forty-eight hours upon potatoes which have been rendered feebly alkaline.

For staining the bacilli in sections, the best solution to employ is composed of 30 c.cm. alcoholic solution of methyl blue in 100 c.cm. of 0·01 per cent. caustic potash (1 part of caustic potash in 10,000 parts of water). A few minutes are required for staining, after which the sections are placed for some seconds in ½ per cent. acetic acid, and then in absolute alcohol. Finally, they are treated with cedar oil and mounted in balsam. The ordinary aniline dye solutions have no effect. Besides the method of Löffler, the *negative* result obtained with Gram's process is to be considered a criterion of the presence of Löffler's diphtheria bacilli.

Various animals exhibit a high degree of sensibility to infection with pure cultures of this micro-organism. When the conjunctiva of young rabbits is slightly abraded and smeared with a pure culture, they soon die, according to Babès, and subcutaneous injections act with fatal effect upon guinea-pigs in two days. Bacilli can be detected at the point of inoculation, but all the remaining organs are free.

Löffler obtained a whitish substance from the cultures by extraction with glycerine and precipitation with alcohol. It is soluble in water, and one or two decigrams when injected hypodermically set up a hæmorrhagic œdema and cutaneous necrosis. Brieger and C. Fränkel state that the poison extracted from the diphtheria bacillus belongs not to the alkaloids but to the toxalbumins.

Numerous cocci are often found in the false membranes of diphtheria, and these were formerly looked upon as the cause of the disease; some of them are, however, streptococci, the rest are saprophytes. According to Löffler and Von Hoffmann, bacilli are frequently detected in the cavity of the mouth and pharynx which morphologically and biologically resemble those of diphtheria, but are devoid of virulence, and these are consequently described as *pseudo-diphtheritic* bacilli. Roux and Yersin, however, believe them to be genuine diphtheria bacilli the virulence of which has become very much attenuated.

Spirillum Milleri.—Miller isolated from a carious tooth a variety of bacterium which appears to be identical with the vibrio described by Finkler and Prior. It consists of jointed rods, sometimes more and sometimes less curved, the figure of which resembles S or O. The spirilla are destitute of motility. Upon the gelatine plate there appear scattered colonies, which speedily liquefy. Thrust-cultures show no air-bubble at the surface, owing to the rapidity with which liquefaction progresses, but there soon forms a tolerably extensive funnel-shaped fluid area, on the bottom of which the masses of bacilli lie (compare fig. 52). A superficial layer forms upon agar. The micro-organism seems to take a prominent part in the softening and liquefaction of the enamel organ which occurs in caries.

Spirochæte dentium (Denticola).—There is found in the cavity of every mouth, according to Miller, and particularly below the edge of the gum, a variety of spirillum which is also met with in the secretion of the nose, and is remarkable for the fact of having pointed ends. This peculiarity is only seen in two of the forms of spirilla discovered up to the present, viz. the *Spirochæte dentium* and the *spirillum of recurrent fever*. Artificial cultivation has not yet been successful.

Vibrio rugula.—Vignal found also an anaerobic micro-organism in the mouth, which, however, can only be preserved in an indifferent gas, such as hydrogen. Yellowish-white globular colonies develop on the gelatine plate, and liquefy outwards from the periphery. In thrust-cultures little white nodules form on the surface as early as the first day, and liquefaction gradually sets in, a white mass lying at the bottom. A white wrinkled coating develops on potatoes and serum, and the latter is speedily liquefied. All cultures generate a gas with a most offensive feculent odour. According to Prazmowski, the spirillum develops terminal spores. Its growth causes curdling in milk, and a very rapid decomposition of albuminoid bodies and of cellulose.

The fungus of thrush (Soorpilz).—The patches known as *thrush*, which occur in the mouths of infants at the breast and of persons greatly reduced by disease, are white in the uncontaminated state, resembling curdled milk, but when they have become impure display various colours. They consist of fibres, conidia-spores, bacteria, epithelial cells,

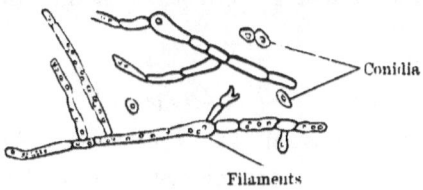

FIG. 90.—FUNGUS OF THRUSH (*Monilia candida*).

and white blood-corpuscles. The thrush-fungus was formerly assigned to the group *Oïdium*, and described as *Oïdium albicans*; but according to Rees, Grawitz and Kehrer it belongs to the yeast fungi, and must therefore be spoken of as *Mycoderma albicans*. Plaut believes it to be identical with a yeast very widely distributed in nature, the *Monilia candida* (fig. 90).

According to Grawitz snow-white colonies develop on the gelatine plate, but do not liquefy the medium. In thrust-cultures white islets form, from which processes radiate out on all sides, so that the culture has the appearance of a fine brush, while filamentous mycelium develops in the deeper part. The fungus grows on potato in a thick white coating, consisting of little masses from the size of a millet seed to that of a lentil strung together in chains. According to Klemperer, guinea-pigs into whose veins a pure culture has been injected perish within from twenty-four to forty-eight hours.

Other bacteria of the mouth.—Besides these, Vignal found in the cavity of the mouth a considerable number of other micro-organisms, such as the *Staphylococcus pyogenes albus* and *aureus*, *Bacillus subtilis*, *Bacillus mesentericus*, and a number of chromogenic bacteria, especially those producing yellow, green, red, and brown pigment. Vignal and Biondi have also found the *Diplococcus pneumoniæ* almost constantly in the cavity of the mouth and the saliva. Rosenbach found *Bacillus saprogenes I.* in white plugs (see p. 175).

The Tympanum

Micro-organisms of the tympanum.—We subjoin the examination of the cavum tympani to that of the mouth, taking the view of Urbantschitsch that it is a diverticulum from the latter.

Netter found the *Streptococcus pyogenes* in diseased conditions of the middle ear, as well as *Staphylococcus pyogenes* and the *Pneumococcus*, but he encountered them not only in middle-ear inflammations in adults, but also in the middle ears of new-born infants. Gradenigo and Penzo very often found the *Micrococcus cereus albus* and *Bacillus lactis aerogenes*, and further the *Micrococcus subflavus*, *Micrococcus candicans*, *Micrococcus flavus tardigradus*, *Micrococcus ureæ liquefaciens*, *Diplococcus lacteus faviformis*, *Bacillus fluorescens*, &c.; so that these observers could find no pathogenic micro-organisms normally present in the middle ear of new-born infants and those at the breast. In *acute otitis media*, Kanthack found chiefly the *Diplo-*

coccus pneumoniæ, and also the *Bacillus pyocyaneus*, amongst others.

The Stomach

Micro-organisms of the stomach.—By examining vomited matters, or by washing out the stomach, some micro-organisms are found in its contents, which are in part derived from the air, having made their way into the region of the pharynx by means of respiration, and passed on thence into the stomach by the act of swallowing; and similarly micro-organisms can be conveyed into the stomach along with water or articles of food, or also from the mouth by means of the saliva. Neither the surface of the mucous membrane lining the stomach, nor its contents, forms, as a rule, a favourable nidus for the various micro-organisms, since the acid contained in it has a destructive action on most of them. Other microbes, on the contrary, resist the acid of the gastric juice, and are also capable of thriving when but little oxygen has access, and hence can grow and maintain themselves for a considerable time in the stomach.

Abelous found certain microbes which are normally present in the stomach. To these *gastric bacteria* belong the *Sarcina ventriculi*, *Bacterium lactis aerogenes*, *Bacillus pyocyaneus a*, *Bacillus subtilis*, *Bacillus amylobacter*, *Bacillus megaterium*, *Bacillus mycoides*, and *Vibrio rugula*. Besides these, other bacteria have been described by this author which are not yet thoroughly known, but amongst which bacilli preponderate. Before Abelous, De Bary had subjected the contents of the stomach to an investigation, and, in addition to different bacteria, discovered the *Oïdium lactis* and the *Leptothrix buccalis*.

It happens in many cases that the acid constituent of the contents of the stomach is reduced to a minimum, or the reaction may even change to alkalinity, so rendering the conditions decidedly more favourable for the growth of micro-organisms, which can pass on into the intestinal canal along with the substances which the stomach contains.

According to Arnold, no micro-organisms are found in the stomach of new-born infants, but on the other hand they can be detected after even twenty-four hours of life.

Sarcina ventriculi.—This sarcina was found by Goodsir in the contents of the stomach, and shows cells resembling cubes with rounded angles and edges, and which are arranged in fairly large-sized packets (see fig. 1). On the gelatine plate rounded yellowish colonies appear in from one to two days, which show diplococci and tetrads under the microscope, but no packets. The gelatine is not liquefied. The thrust-culture shows a growth along the track of the needle and a prominence upon the surface (*nail-culture*). A coating of a yellowish-brown colour, consisting of round islets, appears upon agar. On potato a yellow colour becomes visible round the place of inoculation, followed gradually by a warty deposit, and the inoculated spot itself is chrome-yellow. The envelope surrounding the sarcina when in packet shape distinctly exhibits the cellulose reaction, which consists of a red colour under the action of iodine and sulphuric acid. While the form of the micro-organism is altered when it is grown on the ordinary culture media, the packet figure is retained in cultivations on neutralised infusion of hay, and can be obtained for observation partly from the pellicle and partly from the precipitated masses. If inoculation be effected from a gelatine plate into the infusion of hay, the packet form will again be assumed by the cocci, which have been arranged on the gelatine as diplococci and tetrads. The sarcina thrives better upon a hay infusion to which two per cent. of cane sugar or glucose have been added (fig. 91).

The *Sarcina ventriculi* very readily takes up the aniline

colours, and does not discharge the stain under Gram's process.

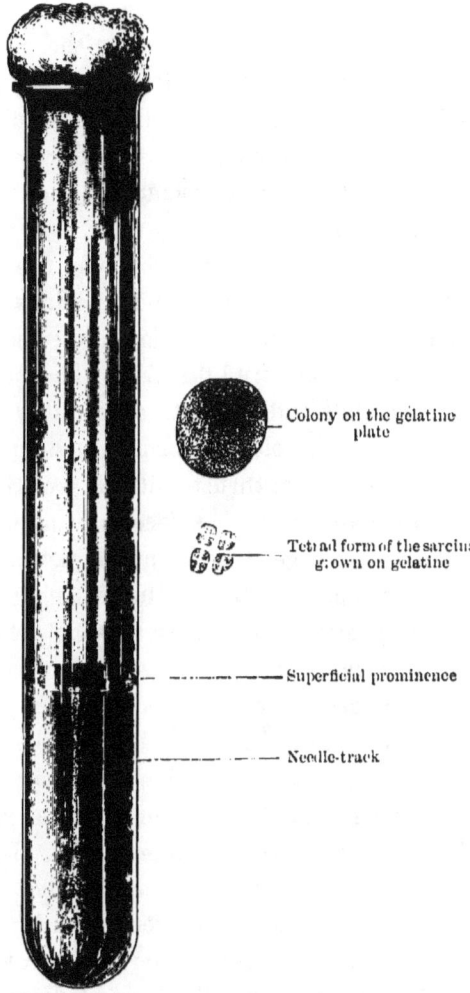

Fig. 91.—Thrust-Culture in Gelatine of Sarcina Ventriculi, four days old.

The growth of this microbe does not affect albuminoid bodies, but it has the property, on the other hand, of precipitating caseine and changing milk sugar into lactic acid.

Micrococcus tetragenus mobilis ventriculi.—Mendoza found in the contents of the stomach motile cocci which tend to arrange themselves in fours, and appear as tablet-cocci. Round colonies form on the gelatine plate, which appear finely granular under a low power and are sharply defined, while on thrust-cultures a dirty-white layer is visible. When the cultures stand for some time they take on a yellowish-brown colour, the *sugar tint*. The gelatine is not liquefied.

Bacterium lactis aerogenes.—In the normal small intestine of young animals and sucklings Escherich found a micro-organism which Abelous detected almost as a rule in the stomachs of adults also. They are short rods with rounded ends and destitute of motility. On the gelatine plate roundish colonies form, which appear yellow in the deeper parts. In thrust-cultures growth takes place along the canal and on the surface a prominence develops (*nail-culture*). The gelatine is not liquefied. A shining layer develops upon agar, having the centre whiter and denser than the periphery. This layer spreads out very rapidly over the surface, and soon appears permeated by bubbles of gas. On serum there forms a greyish moist deposit with irregular edges, which rapidly extends without liquefying the medium. On potato there develops a white coating pervaded by gas-bubbles, which becomes yellowish and fluid. Gas-bubbles appear on other media also, but on gelatine and agar only when glycerine is added. It grows on milk even when oxygen is excluded. Curdling of milk is caused by the growth of the bacterium, and the lactose is very rapidly and energetically converted into grape-sugar. It has no action on albuminoid bodies.

Bacillus indicus.—Koch found in the gastric contents of an Indian ape a fine, very short bacillus with rounded ends, and characterised by a tolerably distinct motility. The gelatine is very rapidly liquefied. Round yellow colonies with bulging edges form on the gelatine plate, being larger on the surface than in depth. In thrust-cultures the upper part of the liquefied mass is coloured brick-red, and on agar there develops a brick-red deposit with white edges. Serum is liquefied, sometimes with, sometimes without, pigmentation. On potato a brick-red deposit forms on the place of inoculation. Pigmentation depends upon the access of air. This bacillus is distinguished from the *Bacillus prodigiosus*, to which it has otherwise considerable resemblance, by its pathogenic action on animals, which it kills with the symptoms of a severe gastro-enteritis.

The Intestine

Intestinal micro-organisms.—Passing from the stomach along with the chyme, the micro-organisms reach the internal surface of the intestine, an occurrence which, however, only becomes possible from the fact that certain of them are capable of preserving their vitality notwithstanding the gastric juice, or that the reaction and other qualities of the juice itself are not such as to exert a prejudicial influence on the microbes. The internal surface of the intestinal mucous membrane contains very many micro-organisms, and seems as a rule to afford a very favourable environment for them, while the contents of the gut, whether liquid or firm, show microbes in such abundance even under normal conditions that of all parts of the body they are richest in such organisms.

The alkaline reaction of the intestinal secretion and the length of time which food takes to pass through the intestinal canal are important factors in favour of the growth of micro-organisms. Duclaux ascribes to them a certain amount of digestive significance, on the ground that their activity is capable of assisting the physiological digestive function, and he believes that the duty has fallen to them of carrying on a '*digestion bactérienne*' by their vital action.

Under normal conditions the microbes lie on the inner surface of the intestine, but in pathological processes the parasites penetrate between the epithelia of the villi, thence into the interior of the villi, and eventually through the intestinal parietes to the peritoneum.

Articles of food which contain bacteria increase the number of colonies obtained from the contents of the gut to a very considerable extent, while the ingestion of sterile food-stuffs or of red wine causes, according to Arnold, the appearance of a striking diminution in the number of germs.

According to Duclaux, only those bacteria can develop luxuriantly in the intestine which have no stringent need of oxygen; but Abelous sets up the hypothesis that although the oxygen necessary for the existence of aerobes occurring in this situation can, in point of fact, reach the intestine with the saliva, the intestinal micro-organisms have the power of accommodating themselves to a deficiency of it.

Most of the bacteria living upon the mucous membrane are saprophytes; the pathogenic varieties are present only in small numbers, and often perish by being overgrown by the former, while many pass away in the fæces. According to Nothnagel, a very large number of yeast-cells are found in the contents of the gut, so much so that Brefeld seeks the normal habitat of yeast in the large intestine.

Micrococcus aerogenes.—Miller found in the digestive tract fairly large non-motile cocci of oval outline, which are distinguished for their marked power of resisting acids, and hence do not lose their vitality in acid digestive juices. Gelatine is slightly liquefied. The colonies on plate-cultures are round, and have on their surface several specks, which under a low power appear sometimes light and sometimes dark. In thrust-cultures the growth takes the form of a nail, and is of a greyish-yellow colour. A yellowish-white deposit develops upon agar ; that on potato shows irregular prominences. Generation of gas takes place in the presence of carbohydrates.

Bacillus putrificus coli.—Bienstock found a micro-organism in the contents of the intestine which has the power of decomposing albuminous substances with formation of ammonia, a process which takes place whether air is excluded or is present. It consists of short motile rods, the growth of which does not liquefy gelatine. A yellowish coating develops upon agar after some time.

Bacillus coprogenes fœtidus.—In the intestine of pigs attacked by *erysipelas*, Schottelius very often found short immotile rods with rounded ends, which play no part in the erysipelas, but can very readily make their way into the blood and neighbouring organs in consequence of the intestinal ulcers. Gelatine is not liquefied, and small pale yellow islets develop in it, which coalesce and form a grey transparent coating. The culture gives off a very unpleasant smell.

Bacterium Zopfi.—Kurt and Flügge discovered in the intestine of chickens actively motile bacilli, the growth of which does not liquefy gelatine. Threads resembling mycelium appear on the plate even in a day, and in thrust-cultures abundant filaments pass out from the needle track, extending in a radial direction, but often crossing each other. No growth takes place on serum. The bacilli thrive best at a temperature between 20° and 30° C. ; higher than this their vitality is reduced.

Bacterium aerogenes, Helicobacterium aerogenes, and Bacillus aerogenes.—These micro-organisms were found by Miller in the intestinal tract. They are motile bacilli which do not liquefy gelatine by their growth, possess the peculiarity of offering resistance to moderately concentrated acids, and hence are able to pass through the stomach without injury to their vitality.

Bacillus dysenteriæ.—Chantemesse and Widal discovered short rods with rounded angles and but scanty power of movement in the contents and walls of the intestines, as well as in the spleen and abdominal glands, in cases of *dysentery*. They take up the aniline colours badly, and do not liquefy gelatine. On the plate there develop at first small white specks, which assume a yellow colour and unite with the neighbouring islets; but in some days the yellow colour vanishes, and the colonies become white and granular. A dry yellow membrane develops on potatoes.[1]

Bacillus of fowl cholera.—*Poultry typhoid* or *fowl cholera* is an epidemic disease accompanied by diarrhœa, often occurring amongst poultry, and which Perroncito first in-investigated more thoroughly. Pasteur describes it as '*choléra des poules*,' and more recently Marchiafava, Celli, and Kitt have been occupied in bacteriological research concerning it.

In the blood, in the capillaries of all the various organs, particularly the spleen and liver, and in especial abundance in the intestinal contents, there are found short plump rods which are somewhat constricted in the centre. A typical attack of the disease can be caused in hens by inoculation with these or by feeding the fowl with them, and other birds also, such as geese and pigeons, are very susceptible of infection, as are, moreover, mice and rabbits. Guinea-pigs, however, prove refractory. The growth of the bacilli

[1] [Many cases of dysentery are, however, believed to be due to the action of a protozoon, named by Lösch *Amœba coli*, by Councilman and Lafleur *Amœba dysenteriæ*, for an account of which the reader is referred to the Appendix.]—Tr.

does not liquefy gelatine, and the colonies on the plate are roundish with uneven edges. When grown in thrust-cultures a delicate film appears on the surface, while a slight whitish-grey coating develops on agar. A dirty yellow fur appears on potato only at incubation temperature, and a shining whitish deposit upon serum.

A characteristic point in the staining of these microbes is that the poles only are coloured by aniline dyes, the centre remaining unstained. They discharge the colour when treated by Gram's process.

Other intestinal bacteria.—Besides the above, the *Bacillus butyricus*, *Vibrio rugula*, and *Bacterium coli commune* are met with in the intestinal canal, as well as the *cholera bacillus*, *typhoid bacillus*, *Vibrio proteus* (*Bacillus Finkler-Prior*), *Bacillus neapolitanus* (*Bacillus Emmerich*), *Proteus vulgaris*, *Vibrio Metschnikoffi*, and in many cases the *thrush-fungus*. [Various protozoa have also been described.]

CHAPTER XII

THE ORGANS AND CAVITIES OF THE BODY AND THEIR
CONTENTS (*continued*)

III. THE FÆCES AND URINE

The Fæces

Composition and modes of examining.—The fæces must at this point be dealt with as a supplement to the study of the intestine, and they are very rich in micro-organisms. When the stools are examined we find constituents derived from the vegetable matters eaten, and remnants of undigested pieces of animal food, especially larger or smaller pieces of striped muscle fibre, which show the transverse striation distinctly. Fat often occurs in normal stools in the form of droplets, and, in addition to these constituents, epithelial cells are also found (usually cylindrical), red and white blood-corpuscles, masses of detritus, crystals, and under pathological conditions still other appearances. Certain varieties of bacteria can be detected in the meconium of new-born infants, as well as in the evacuations of adults. According to Nothnagel, micrococci are found in preponderating numbers in firm, and bacilli in fluid, stools.

In the examination of fæces, forms are found almost constantly which stain blue with tincture of iodine, and which Nothnagel considers to be identical with the *Clostridium butyricum*. Other micro-organisms also assume a blue colour under application of tincture of iodine; for example, Von Jaksch found rods in the fæces which recall the appearance of *Leptothrix buccalis*. The number of microbes which stain with iodine is found to be particularly large in intestinal catarrhs. Fischer succeeded in growing the Leptothrix fibres artificially. Furthermore, the different micro-organisms which live in the intestinal tract are also to be found in the fæces.

When observing the fæces it is important to examine a small quantity in the hanging drop, when, besides the micro-organisms, a copious quantity of desquamated cylindrical epithelial cells are found. Even in such an investigation as this it is very often seen that, in the case of the majority of the different micro-organisms described up to this point, pure cultures of one or other are present, the other microbes either having been suppressed altogether or being present in such small numbers that no attention need be paid to them. Hence it happens, for example, that, besides the cylindrical epithelium, nothing but actively motile cholera bacilli are encountered on examining the rice-water stools of that disease.

Yeast cells are constantly found in the fæces, especially in the acid stools of children (*summer diarrhœa*), and in acute catarrh of the small intestine in adults. These cells stain brown with iodine, which Von Jaksch connects with the glycogen they contain.

Bacillus subtilis appears with tolerable frequency in normal as well as pathological alvine evacuations; and besides these, in diseased conditions of the intestines the respective micro-organisms are met with in the fæces —for instance, the *bacilli of cholera, typhoid fever*, and *tuberculosis*. Netter was also able to isolate the *Staphylococcus pyogenes* from the fæces.[1]

---

[1] [Lambl, Lösch, Koch, Pfeiffer, Kartulis, and others have also described protozoa occurring in the fæces in health as well as under various pathological conditions.]—Tr.

Bacillus subtiliformis.—Bienstock found, as a constant inhabitant of human fæces, a bacillus the morphological characters of which strongly resemble those of *Bacillus subtilis*, except that motility is wanting and the rods are always connected in long filaments. A luxuriant fatty coating of a whitish-yellow colour develops upon agar. When spores form, the rods are distended in the centre to a spindle shape.

Bacillus albuminis.—The same observer very often encountered a micro-organism in fæces, which possesses the power of energetically decomposing albumin, and to which he gave, in consequence, the name *Bacillus albuminis*. The rods are fairly long and show a marked motility, that part of the bacillus from which the spore divides off being always foremost during movement. A whitish layer, with the lustre of mother-of-pearl, develops on agar.

Bacillus cavicida.—Brieger isolated from fæces and putrid substances a bacillus which has the property of decomposing saccharine solutions and generating propionic acid. The rods are short, being but twice as long as their diameter, and liquefy gelatine into a viscid fluid. On plates there develop white colonies with beautiful concentric rings resembling the scales upon the back of a tortoise. A dirty yellow deposit forms on potatoes and on serum. Subcutaneous injections have an extremely poisonous action upon guinea-pigs, often causing death even within a few hours.

Micrococcus tetragenus concentricus.—The author found in the liquid evacuations of a person suffering from gastric dilatation cocci which were commonly united in tetrad form. The elements are small, round and motile, stain readily with all the different aniline colours, and thrive upon the various nutrient media at present used in bacteriological research. Gelatine is not liquefied. On the plate several round islets appear, which when inoculation has been extensive coalesce with one another, and by confluence of several

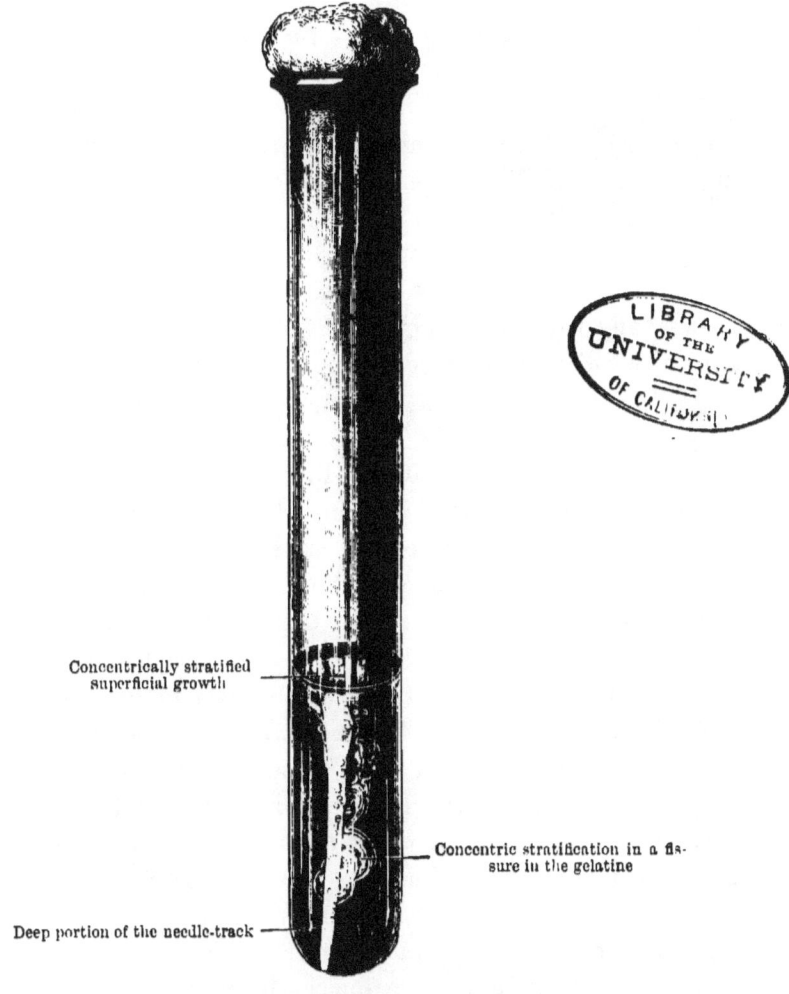

Concentrically stratified superficial growth

Concentric stratification in a fissure in the gelatine

Deep portion of the needle-track

FIG. 92.—THRUST-CULTURE IN GELATINE OF MICROCOCCUS TETRAGENUS CONCENTRICUS.

a larger roundish compound colony is formed, resembling a holoblastic ovum in process of segmentation. Thrust-cultures show a superficial deposit possessing concentric rings, which may be described as *diurnal rings*, since a difference exists between the growth in the dark and that in daylight, which expresses itself in this form (fig. 92). If the culture be kept in the dark the concentric circles on the upper surface disappear. On agar there appears a deposit showing scroll-like tracings, and on potato growth takes place on the site of inoculation (fig. 93.) The microbe cannot be found to possess any pathogenic properties, subcutaneous inoculations upon mice having produced no results.

FIG. 93.—POTATO CULTURE OF MICROCOCCUS TETRAGENUS CONCENTRICUS.

The Urine.

Micro-organisms of the urine.—In the normal state no micro-organisms are found in urine when freshly passed; those which appear in it reach it from without, or are derived from diseased conditions of the urinary passages. If, however, the urine stands for some time until alkaline fermentation sets in, a multitude of yeasts, moulds, cocci, and bacilli make their appearance, which, reaching the urine probably from the air, and finding there a favourable environment, set up a fermentation of the urinary sugar and cause a splitting up of the urea into ammonium carbonate; in fact, an entire series of processes taking place in the urine are dependent on the vital activity of micro-organisms.

A true *bacteriuria* has also been described in different quarters (Roberts, Schottelius, Fischer), which appears to be of a morbid nature: in pathological processes in the different organs, but especially when bacteria get into the blood and therewith also into the renal circulation, bacilli may penetrate into the urine during its secretion and be expelled with it.

In a case of *endocarditis*, Weichselbaum found the specific micro-organisms in the urine, and Von Jaksch detected in *erysipelas* abundant streptococci which he identified with the *Streptococcus pyogenes*; while in *typhoid fever* Neumann found the corresponding micro-organisms in the urine, and *glanders bacilli* and *tubercle bacilli* have also been discovered in it. In other diseases, too, the respective bacteria are found in the urine—for instance, the *bacillus of anthrax* or the *spirillum of relapsing fever*. In gonorrhœal processes *gonococci* are found in it, and in suppurative processes considerable numbers of staphylococci; and tubercle bacilli appear in S-shaped groups (see fig. 81) in tubercular ulcerative disease of the urinary passages. The process of centrifuging is especially to be recommended when examining the urine for micro-organisms.

Bacteriuria may also be caused by the conveyance of micro-organisms into the urethra in the introduction of instruments, when they may set up a decomposition of the urine in the interior of the bladder.

A number of different micrococci found in the air, in water, and in the soil are also met with in the urine; these include the *Micrococcus ureæ* and various species of sarcina.

The majority of the micro-organisms occurring in urine are bacilli.

Yeasts and moulds in urine.—The occurrence of any considerable number of yeast-cells is in all probability dependent upon the urine being rich in sugar. Moulds, too, only appear in saccharine urine, and then not until the alcoholic fermentation of the sugar has come to an end.

Urobacteria.—A number of bacteria have the property of converting urea into ammonium carbonate. That most frequently found in the normal urine during fermentation is the *Micrococcus ureæ* (Pasteur, Van Tieghem, Leube), which is a constant inhabitant of the atmosphere, and has consequently been already described in the chapter on the 'Bacteriological Analysis of Air.' It is usually found in longish chains upon the surface of fermenting urine (see fig. 34, p. 107).

Lundstroem found that two kinds of staphylococcus are also endowed with this property, viz. the **Staphylococcus ureæ candidus** and the **Staphylococcus ureæ liquefaciens**. The former develops shining white deposits upon the gelatine without liquefying it, the latter liquefies it.

Miquel encountered amongst the water bacteria numerous varieties possessing the same power. They are for the most part bacilli, only a few being micrococci; amongst the latter are found three which liquefy gelatine. According to Leube, a variety of sarcina is also to be included with these micro-organisms.

The **Micrococcus ureæ liquefaciens** described by the last-named observer forms small chains and liquefies gelatine slowly.

Leube isolated a **Bacillus ureæ** from ammoniacal urine, consisting of short stout rods, the growth of which does not liquefy gelatine. The colonies on the gelatine plate are at first very transparent; their coalescence gives the gelatine on that spot the appearance of a slab of ground glass. The margins are indented in consequence of irregularity of growth. In thrust-cultures also grey processes appear, extending out from the needle-track, and these grow at room temperature.

Of other micro-organisms which decompose urea the **Urobacillus Freudenreichii** and the **Urobacillus Maddoxii** should also be mentioned. The former grows best on gelatine at 20° C., forming a milk-white coating on the surface, while deeper a cavity filled with a turbid stringy fluid develops. Liquefaction of the gelatine progresses slowly, and the liquefied region gradually clears, a white slimy mass with the smell of ammonia settling down into the deeper part. If some urine be added to the gelatine a coronet of very fine crystals forms after some days around the white colonies. It also grows excellently upon bouillon and upon urine.

The *Urobacillus Maddoxii* renders urine viscid and stringy. Gelatine is not liquefied, bouillon is very quickly rendered turbid, and in this case also crystals form round the cultures in gelatine mixed with urine.

Micrococcus ochroleucus.—Prove found cocci in normal urine which are possessed of a motility particularly marked when they are united in chains, and Legrain met with them also in the pus of *urethritis*. They are distinguished by the formation of endogenous spores, which progresses best at 36° C. The gelatine is not liquefied. After even one day colonies with raised edges appear upon the plate, and these later on become yellow and push out processes. In thrust-cultures a sulphur-yellow pigment forms upon the surface, which is soluble in alcohol and is destroyed by acids. A dirty white, creamy layer develops on agar, in which the inoculated streak is prominent as a yellow stripe. On potato there forms a warty elevation of a yellow colour. All cultures diffuse an intense sulphurous odour.

Streptococcus giganteus urethræ.—Lustgarten and Mannaberg have discovered in normal urine and in the human urethra large cocci arranged in rows, which form wavy lines and often dense coils. No growth takes place on gelatine, and even on agar it is slow, and best, moreover, at incubation temperature.

Besides these, the same observers described four varieties of bacilli and seven of cocci as constant inhabitants of the urethra, amongst them the *Staphylococcus pyogenes aureus* and *Micrococcus subflavus*.

Bacterium sulphureum.—Rosenheim found different bacteria in urine which form sulphuretted hydrogen therein and do not liquefy gelatine.

Holschewnikoff isolated a micro-organism from mud, which generates the same gas, and is also, although rarely, to be met with in urine. It consists of fine rods with rounded ends, and endowed with a slow motility. Their development causes liquefaction of the gelatine. Upon plate-cultures there form small punctate colonies, which sink inwards in funnel form during liquefaction. In thrust-cultures the superficial colonies are white, the deep reddish-brown. A viscid grey layer forms rapidly on agar; but on potato development takes place only when air is excluded, the growth being reddish-brown.

Bacillus septicus vesicæ.—In persons suffering from *cystitis* and *pyelonephritis* Clado found, amongst numerous other microbes, an organism showing motile rods, mostly isolated, and which develop ovoid spores. They take up aniline stains readily, and do not discharge their colour under Gram's process. Gelatine is not liquefied. Upon the plate there develop punctiform colonies, which, however, do not go beyond a pin's head in size. In thrust-cultures the deeply-seated colonies appear larger than the superficial, and the latter unite to form a delicate opalescent film. On agar there develops a delicate coating, and upon this minute circular shining milk-white colonies grow. Both gelatine and agar quickly become alkaline. Growth in bouillon is particularly abundant, and a dry, light-brown layer develops upon potato.

Urobacillus liquefaciens.—Schnitzler has found a bacillus in *cystitis* to which the above name has been given, and which is apparently identical with the *Urobacillus liquefaciens septicus* discovered by Krogius in *cystitis* and *pyelonephritis*. It consists of short motile rods rounded at the ends, which give up the stain when treated by Gram's method. Gelatine is rapidly liquefied. The plate-culture shows in the centre of the liquefied colony a nodule of the size of hemp-seed, with fringed edges. There develops a thick greyish-white unwrinkled coating upon agar even in one day. On potato a brownish-yellow layer forms. The cultures generate ammonia and smell like decomposed urine. This bacillus seems to be nearly akin to the *Proteus vulgaris* (see p. 175).

The *Staphylococcus pyogenes albus* and *aureus* are also very often found in *cystitis*.

CHAPTER XIII

THE ORGANS AND CAVITIES OF THE BODY AND THEIR
CONTENTS (*continued*)

IV. BACTERIOLOGICAL EXAMINATION OF THE RESPIRATORY
TRACT; AND (V.) OF THE BLOOD

The Nasal Secretion

Micro-organisms in the nasal secretion.—The micro-organisms occurring in the vicinity of men and animals may very easily penetrate into the respiratory passages along with the current of air. They first reach the cavities of the nose, and find in the alkaline viscid mucus there a medium possessing the conditions most suitable for their development. The secretion of the nose contains epithelial cells, both pavement and ciliated, leucocytes, and even in the normal state a considerable number of micro-organisms—cocci, bacilli, and spirilla. Under pathological conditions the nasal mucus becomes thin, and more pus-corpuscles are found, as well as the characteristic micro-organisms in definite local morbid conditions, the kinds of microbe depending upon the variety of disease.

Micrococcus cumulatus tenuis.—Von Besser isolated from the normal nasal mucus rather large non-motile cocci, the growth of which does not liquefy gelatine. Upon agar a development of raised colonies appears, showing a thin zone with ragged edges around a central nucleus. Potatoes are not favourable for its growth.

Mirococcus tetragenus subflavus shows non-motile cocci

arranged in fours, and was discovered by Von Besser in the normal mucus of the nose. No growth takes place upon gelatine, but it develops well on agar and potato. The older cultures have a yellowish-brown colour.

Micrococcus nasalis.—Hack encountered in the nasopharyngeal cavity motile diplococci, the growth of which does not liquefy gelatine. The islets on the gelatine plate show a small undulating excavation in the centre, and in individual colonies it is easy to recognise a radiating arrangement or coiled form, frequently with a rayed margin. A concentric stratification of the superficial layer is often seen in thrust-cultures, together with development of nodules along the thrust-canal. A greyish-white shining deposit forms upon agar and potatoes.

Diplococcus coryzæ.—This diplococcus was described by Hajek, who found it in the nasal secretion. It consists of cocci which are somewhat elongated, so as to look like short bacilli. Gelatine is not liquefied by them. Flat superficial-lying colonies develop on the gelatine plate, and on thrust-cultures a deposit forms, which is at first raised but becomes continually more and more flattened out, a fact which serves to distinguish the culture of *Diplococcus coryzæ* even by the naked eye from Friedländer's *Pneumobacillus*, since no alteration takes place in the nail-culture formed by the latter. A coating develops upon agar.

This diplococcus was at first regarded as the exciting cause of those changes in the mucous membrane which are characteristic of *coryza*. Experiments on animals, however, have yielded results which are negative in this respect.

Staphylococcus cereus aureus.—This microbe resembles the *Staphylococcus cereus flavus*, only differing from it in the orange-red colour of its colonies, which is particularly marked in thrust-cultures. It was discovered by Von Schrötter and F. Winkler in the author's Institute, in the thin secretion of a cold in the head, and in association with the *Staphylococcus cereus flavus* (see fig. 76).

Bacillus fœtidus ozænæ.—Hajek found in the secretion of patients suffering from *ozæna* short actively motile rods, during the growth of which the gelatine is liquefied. Roundish colonies form at first upon the gelatine plate, which appear sunken like pock-marks while liquefaction is beginning, but later become confluent and liquefy the entire gelatine. In thrust-cultures the liquefaction shows itself both superficially and along the track of the needle, and if the culture is kept at 37° C. a strong odour of putrefaction is developed, which is not the case at the ordinary temperature. Upon agar a superficial coating forms, and in like manner becomes foul-smelling at incubation temperature. A brownish deposit forms on potato. If animals be injected subcutaneously with the cultures a violent inflammation results, in the course of which the tissues become necrosed. The bacilli completely discharge their colour when treated by Gram's method. Hajek particularly recommends staining with alkaline methyl blue to which some aniline water has been added. Diluted alcoholic solutions stain the bacilli only very slightly.

Bacillus striatus albus et flavus.—Both bacilli were met with by Von Besser in the normal nasal mucus, but the *Bacillus striatus albus* is very rare. Gelatine is not liquefied. A fairly good growth appears on the different nutrient media. Upon potato a streaky layer forms. *Bacillus striatus flavus* develops a sulphur-yellow pigment.

Bacillus of rhinoscleroma.—The *bacilli of rhinoscleroma* are short rods rounded at the ends, which are devoid of power of automatic movement and are enclosed in capsules. They are found in rhinoscleroma in the tissue of the tumours and in the juice from them, and are believed to be the cause of this form of disease. Von Frisch, Paltauf, Von Eiselsberg,

Dittrich, Cornil, Alvarez and several other investigators have made researches concerning this micro-organism.

The best solution to use for staining sections of the rhinoscleroma tumours when it is desired to examine them for the presence of the bacilli, is the methyl blue or methyl violet solution of Löffler. The sections remain one or two days in the stain, are then washed in water containing iodine, and finally decolorised for a rather long time (two or three days) in absolute alcohol. The bacilli do not lose their colour under Gram's process. If they be stained with gentian violet in aniline water, washed in water containing acetic acid, and double-stained with carbolic fuchsine, the capsules round the bacilli appear coloured; safranine may be used with advantage for double-staining instead of the carbolic fuchsine.

Gelatine is not liquefied. Roundish colonies appear on the gelatine plate, and thrust-cultures take the nail form, but do not attain any very extensive development. Upon agar a widespread milk-white coating appears even in half a day, on potato there likewise forms a creamy mass which gradually spreads out from the site of inoculation, and on blood-serum also a white coat develops. In all these methods of cultivation the capsule is retained even in the later stages of development, but it is lost in bouillon cultures. Inoculations do not produce rhinoscleroma, even when the injections are made into the nasal mucous membrane, but according to Stepanow tumours of granulation-tissue develop when one is made into the eye of a guinea-pig, and these tumours are rich in bacilli. Injections into the pleura of these animals resulted in death.

The bacillus of rhinoscleroma seems to be at least closely akin to Friedländer's *Pneumobacillus*, but is less virulent. The points of difference with reference to their morphology which might be cited would perhaps be the greater transparency of gelatine cultures of the former, and the persistence of capsules on most of the culture media.

Bacillus capsulatus mucosus.—At the Institute of Professor Klemensiewicz in Graz, Fasching discovered, in crusts removed from the naso-pharynx of a case in which this cavity was diseased, a micro-organism with short and rather thick rods lying singly or in pairs or fours in a common enveloping capsule. The bacilli give up the dye under Gram's process. It is easy to bring out the capsule successfully as a delicate rose-tinted area by protracted staining of the prepared cover-glass in fuchsine, or by slightly warming after covering it with the stain. The bacilli are destitute of motility, and do not liquefy gelatine. On the plate colonies develop which have the appearance of drops of mucus of the size of pins' heads. In thrust-cultures there forms a typical nail-like figure, with active generation of gas, and upon agar and serum a thick moist creamy layer develops. A moist, viscid, white coating occurs on potato. Subcutaneous injection causes a genuine septicæmia in white mice. Fasching also found this micro-organism in the sputum of a phthisical patient.

Vibrio nasalis.—Both in the nasal mucus and in the buccal cavity rods of considerable size are found which possess no power of automatic motion, and are arranged as vibrios. They were cultivated pure by Weibel. Gelatine is not liquefied, and the growth on the plate leads very slowly to the formation of round islets. In thrust-cultures a delicate white streak develops along the thrust-canal, resembling a string of mucus. The culture on agar is less transparent and thicker, and spirilla are found in it which show a large number of bends (over thirty). No growth takes place on potato. When treated by Gram's process the spirilla lose their colour.

Other nasal bacteria.—Reimann isolated a great number

of bacteria from the nasal mucus, amongst which also the *Spirochæte dentium* is often to be met with.

In tubercular diseases of the cavity of the nose the pathognomonic *tubercle bacilli* are found in the nasal mucus, in glanders ulcers the *Bacillus mallei*, and *thrush fungus* and *moulds* frequently occur in the secretion. [See p. 195.]

The Respiratory Passages

Micro-organisms of the respiratory passages.—Micro-organisms may also be very easily conveyed from the air into the bronchi by respiration. Straus and Dubreuil have ascertained that the expired air is almost free from organisms, and even when germs do occur the proportion of those in expired to those in inspired air is very small (1 : 600). *Tubercle bacilli* produce infection with greater ease, as Cadéac and Molet found, when the particles of dust to which they adhere are damp with aqueous vapour, than when they reach the lungs in a dry state. Owing to the discovery in recent years of a series of micro-organisms which excite morbid processes in the lungs, it is of very great importance to acquire a more intimate knowledge of the microbes occurring there, and this is most conveniently done by a thorough examination of the sputum. Certainly there are found in the sputum admixtures of matter from the naso-pharynx and the buccal and nasal cavities; still, it is possible to exclude all the elements coming from other cavities, and to retain only those derived from the lungs.

Pansini constantly found different varieties of streptococci in the sputum both of healthy and diseased individuals, and next to these in frequency, various kinds of sarcina. A yellow, reddish, or green colour in sputum is caused by chromogenic bacteria. Pansini derives the yellow and reddish colour from the vital activity of *Bacillus aureus*,

Sarcina lutea, and *Sarcina aurantiaca*, while the green colour is caused by the *Bacillus pyocyaneus* and *Bacillus fluorescens non-liquefaciens*. Virchow and Lichtheim not rarely found *moulds* in the sputum, and amongst these particularly *Aspergillus fumigatus*. The *fungus of thrush* and *yeast cells* may also be frequently met with, and moreover, Fasching found here his *Bacillus capsulatus mucosus*.

Sarcina pulmonum.—Even in earlier years different authors have described sarcinæ in the sputum which they considered identical with *Sarcina ventriculi*. By the researches of Hauser, however, it was brought out that a special variety has here to be dealt with, distinguished at once from the last-named by its smaller size. No pathogenic properties can be ascribed to it, but it possesses the faculty of energetically decomposing urine, a power shared by it with a sarcina discovered in that excretion by Leube. Gelatine is not liquefied, and on the plate small white colonies develop which appear indented at the margin and concentrically stratified during their growth. A superficial moist coating forms in thrust-cultures. Upon potatoes the growth is very scanty.

Sarcina aurea.—Macé isolated from the lung of a patient, who had died of *pneumonia with purulent pleuritis*, a variety of sarcina, the elements of which possess a very lively oscillating motility. Some of them stain bluish-violet with iodine and sulphuric acid, a reaction indicating the presence of cellulose or starch in the sarcina. They develop a beautiful golden-yellow pigment which is soluble in absolute alcohol, and liquefy gelatine with tolerable rapidity. In thrust-cultures a thick friable membrane, which very readily falls apart into separate pieces, develops upon the surface of the funnel-shaped area of liquefaction; on agar there forms a thick streak with warty surface and of a shining golden-yellow colour, while a thick golden-yellow layer grows on potato.

Diplococcus pneumoniæ.—Klebs and Eberth long ago pointed out the presence of cocci in *croupous pneumonia*, while A. Fränkel and Weichselbaum have studied them with the help of the modern methods of research, and ascertained that the cause of this affection is a micrococcus which Weichselbaum detected in ninety-one cases out of a hundred.

These cocci are round or sometimes elongated, and possess a jelly-like envelope of varying thickness. They are very often found arranged in pairs, and it is not unusual for several such diplococci to be connected in rows, one behind the other, and enclosed in a common capsule (fig. 94). Motility is absent.

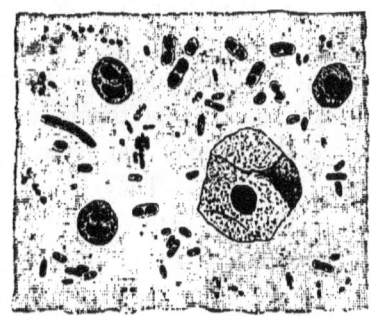

FIG. 94.—MICROBES OF PNEUMONIA.
(After Jaksch.)

These cocci occur in croupous pneumonia not only in the sputum and in the diseased lung, but also in the blood, and they are met with in different other diseases. According to Weichselbaum they are found in exudations in the cavum tympani and in the ethmoidal labyrinth, and Foà and Bordoni believe them to be the sole cause of *cerebro-spinal meningitis*. Klein, Biondi, A. Fränkel and Miller found them also in the buccal and naso-pharyngeal cavities of healthy individuals, so that they can exist as it were in the portals of entrance to the respiratory system. They are identical with the microbes of *sputum septicæmia*. Emmerich found them in the dust of a room occupied by pneumonic patients.

Their growth begins to make good progress at a temperature of 24° C., and gives the best results at incubation heat. On the gelatine plate there form roundish colonies of medium size, and in thrust-cultures little white granules appear along the needle-track, and a slight transparent prominence on the surface. The gelatine is not liquefied. A thin transparent film forms upon agar, and a thin transparent slimy coating upon serum. The cultures do not last very long, and lose their virulence if kept at a temperature of 42° C. even for one or two days. The same result is also attained, according to Fränkel, by cultivation in milk. In order to preserve the virulence of the cocci, they must be inoculated from time to time upon animals.

The diplococci cultivated upon artificial nutrient media show no capsules, and after losing them they become regularly round and range themselves in chains, in consequence of which they have been described by Gamaleia as *Streptococcus lanceolatus*. They exhibit capsules in the blood of animals infected with the cultures.

The cocci stain in dilute alcoholic solutions of the aniline dyes, and can easily be displayed in preparations coloured by Gram's method, differing therein from the *Pneumobacillus* of Friedländer. The capsules remain unstained by the ordinary methods: to colour them Ribbert employs a hot saturated solution of the capsules of dahlia violet in 100 parts water, 50 parts alcohol, and $12\frac{1}{2}$ parts acetic acid. Staining takes place very rapidly in this solution, and moreover it is necessary to wash in water only for a short time. The cocci appear dark blue, the capsules light blue.

For transmission of pure cultures those on bouillon are the most suitable, and of these one or two cubic centimeters are used as a hypodermic injection. The cocci with their capsules are found in the blood and organs, but subcutaneous injections fail to set up inflammatory symptoms in the lungs. When the pleura is infected, however, inflammations of it and the lungs do occur; and when pure cultures are introduced into the trachea of rabbits a pneumonia with all the characteristic symptoms follows.

Pneumobacillus Friedlaenderi.—Friedländer found a micro-organism in the expectoration and in the lung-tissue, the elements of which are rods of different sizes, lying singly or joined together in pairs or bands. They possess a capsule in the form of a transparent surrounding area, but this is wanting in artificial cultures. The bacilli are immotile. In contrast to the *Diplococcus pneumoniæ*, the pneumobacilli discharge the dye under Gram's process. They also grow at a lower temperature than the former.

Gelatine is not liquefied. Upon the plate there appear roundish, sharply-defined colonies of granular texture, and in thrust-cultures a thick porcelain-like prominence forms on the surface, and the growth rapidly advances along the thrust, so that there appears the distinct figure of a nail—the '*nail-culture*' (fig. 95). Older cultures become brownish. Upon agar a dense deposit forms, and a thick yellowish moist coating upon potato. Bubbles of gas are often seen in the cultures. Infections of mice by the hypodermic method speedily result in death, but guinea-pigs are less susceptible. Sub-pleural or intra-pulmonary injections set up a pneumonia. *Croupous pneumonia* seems, however, to be due to the pneumobacillus only in a small proportion of cases (9 times in 100 cases), according to Weichselbaum and C. Fränkel.

Micrococcus tetragenus.—Koch and Gaffky found in the contents of a tubercular cavity small immotile cocci, which were, as a rule, united with one another in fours and surrounded by a capsule. They are also found in the sputum, and are very frequent, according to Kar-

T

linski, in dental abscesses. Gelatine is not liquefied, and white punctate colonies form on the plate and display a

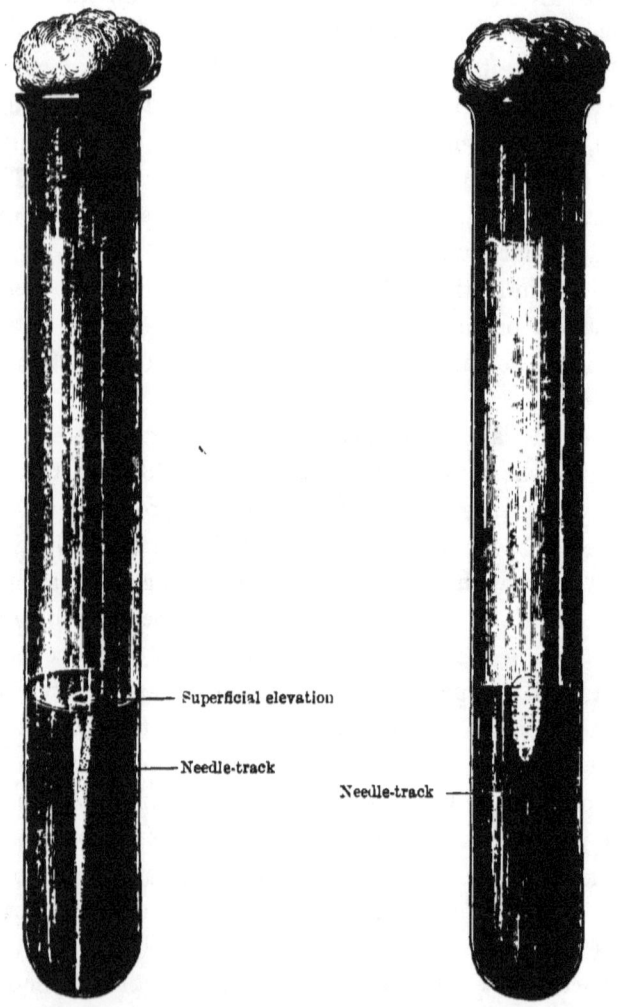

Fig. 95.—Thrust-Culture in Gelatine of Friedländer's Pneumobacillus ('Nail-Culture').

Fig. 96.—Thrust-Culture in Gelatine of Micrococcus Tetragenus. (After Macé.)

shining, porcelain-like appearance on the surface of the medium. Isolated sharply-defined colonies appear along the canal in thrust-cultures, lying one above the other like a pile of discs, the most superficial of which projects in a hemispherical form above the surface (fig. 96). Upon agar and serum, also, round well-defined colonies develop along the inoculated streak, and upon potato a slimy coating forms. White mice die within four weeks after infection.

Bacillus aureus.—The elements of this bacillus are short rods showing but slight motility. They were found in water by Adametz, and also upon the human skin in some forms of *eczema* by Unna and Tommasoli. Gelatine is not liquefied. Upon the plate there develop punctiform colonies, which assume a yellow colour and become uneven, and in the thrust-culture a dark yellow deposit forms upon the surface. Their growth on potato takes the form of shining hemispheres, which coalesce and assume a colour varying from dark yellow to brownish-red.

Tubercle bacillus and Actinomyces.—The micro-organism to which Koch gave the name of *tubercle bacillus* is constantly found in the sputum of phthisical persons and in the contents of cavities; it has already been described in the chapter on the 'Bacteriological Analysis of Pus' (p. 206). The *actinomyces fungus*, which often occurs in sputum, has also already been described in the same chapter (p. 196).

Bacillus tussis convulsivæ.—Afanassiew found a micro-organism, which he described under this name, in the sputum of persons suffering from *whooping-cough*. The rods are short and actively motile, and best admit of being cultivated at incubation temperature. Gelatine is not liquefied by them. Rounded brown colonies form on the plate and a superficial coating in thrust-cultures, while there develops on agar a thick grey, and on potato a thick brown, deposit.

Bacillus pneumosepticus.—In the respiratory organs, and

in various other tissues of an individual who had died of *septic pneumonia*, Babès found short immotile rods which readily take up aniline dyes, but discharge them again under Gram's process. The gelatine is not liquefied by them. Extensive flat white islets develop on the plate, and in thrust-cultures growth takes place along the entire needle-track and diffuses an unpleasant, mawkish odour. Upon agar indistinct colonies appear, which coalesce and form a superficial film. A similar thick moist coating is also found upon potato. Experimental inoculations on mice, rabbits, and guinea-pigs speedily cause death. It is not unusual for the bacilli to be found enclosed in the leucocytes also. Virulence is lost after repeated transmissions to artificial nutrient media.[1]

V. Bacteriological Examination of the Blood

Micro-organisms in the blood.—In treating of the various micro-organisms it has already been pointed out in a general way that they are found not only in the several organs, but also in the blood, and hence examination of the latter must be undertaken both in microscopic preparations and by the methods of cultivation.

We find in the blood moulds, yeasts, and bacteria, and it has been mentioned already in an earlier part, that staphylococci, streptococci, *tubercle bacilli* (Weichselbaum), and the *bacilli of typhoid fever, anthrax, glanders*, &c., may be recognised here also. Besides these, however, there are other organisms to be cited which occur *par excellence* in the blood.

Methods of examination.—When it is desired to examine human blood, the ball of the finger must first be carefully cleansed with sublimate, alcohol, and ether, after which

---

[1] [The occurrence of *Amœbæ* in the sputum has also been described in cases of *dysenteric abscess*. See Appendix.]—Tr.

the blood is obtained by pricking it. The first drop having been wiped away with a sterilised platinum needle, the second is taken off with a disinfected cover-glass, spread out by rubbing with another cover-glass (also disinfected), dried, and heated to 120° C. approximately. Disinfection of the cover-glasses is best done in corrosive sublimate, from which they are transferred to alcohol and ether with a sterilised forceps, and then taken out and rapidly dried in the air. The basic aniline dyes, commonly used for bacteriological examination, are employed in aqueous solutions, with the occasional addition of alcohol, glycerine, or acetic acid to the fluid. The stains are allowed to act for some minutes and then rinsed off with distilled water; the preparation is dried and Canada balsam applied to it, or if it is not wished to keep the specimen for any length of time, it is mounted temporarily in oil of cloves, origanum, or cedar. Löffler's alkaline solution of methyl blue [pp. 86, 241], and Gram's method with its modifications (p. 76), are also much used.

According to Günther's method, the preparations of blood after being dried and fixed are rinsed in dilute solution of acetic acid (1 to 5 per cent.), by which means the hæmoglobin is extracted from the corpuscles and a great part of the plasma washed from the glass, without thereby impairing the adhesion of the bacteria. If now the sections are again dried they may be stained after the ordinary methods, and a tolerably isolated coloration of the bacteria so effected. The blood-corpuscles are no longer seen save as mere shadows, and do not now interfere with the appearance of the stained bacteria.

The rinsing away of the plasma with acetic acid does not succeed if the layer of blood be already too long dried upon the cover-glass. Günther accordingly treats dried-up layers with a 2 to 5 per cent. aqueous solution of pepsine, when the plasma is peptonised in a short time, and the bacteria remain well preserved.

Influenza bacillus.—Babès, Canon, Pfeiffer, and Kitasato believe the exciting cause of *influenza* to be very diminutive rods which can be detected in the blood of persons suffering from the disease. They lie in part within the white corpuscles, and often appear ranked with one another in chains of three or four. In order to detect them, Canon lays the cover-glass preparation, after it has been dried by the air, in alcohol for five minutes, and stains it for from three to six hours in Chenzynski's solution of methyl blue and eosine, rinses it in water, allows it to dry, and mounts in Canada balsam. By this process the red blood-corpuscles are

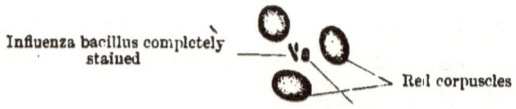

Fig. 97.—Influenza Bacilli in Human Blood. (From an original preparation by Canon.)

stained red, the white corpuscles and bacilli blue (fig. 97). *Chenzynski's solution* consists of 40 grams concentrated aqueous solution of methyl blue, 20 grams of a half per cent. solution of eosine in 70 per cent. alcohol, and 40 grams water.

Upon glycerine agar solidified in a slanting position there develop at incubation temperature very small droplets of the transparency of water,—so small, in fact, as to be scarcely perceptible unless magnified by a lens,—which have no tendency to coalesce. Scanty white turbidities form in bouillon in the first twenty-four hours, and sink to the bottom, leaving the supernatant liquid clear.

Bacillus endocarditis capsulatus.—In a thrombus in the cardiac auricle of a person who had died of *endocarditis*,

Weichselbaum found bacilli enclosed in a capsule and resembling the *Pneumobacillus*. Gelatine is not liquefied, and the thrust-culture shows a shining deposit resembling stearine, while a greyish-white coat develops upon agar. The bacilli discharge their colour when treated by Gram's method. Subcutaneous injections cause the death of the animals experimented on, and if the aortic valves be injured, endocarditis sets in after infection.

Bacillus of swine erysipelas.—One of the most fatal diseases affecting pigs is *swine erysipelas*, since the majority of the animals attacked fall victims to this infectious disorder, often even in a few hours. Feverish symptoms set in, and at the same time there appear over the neck, abdomen, and breast spots which are at first red, but later assume a brown colour. The cause of this disease is considered by Pasteur, Thuillier, Löffler, Schottelius, and Schütz to be a microorganism found in the blood and juice of various organs, particularly the spleen, and consisting of small rods endowed with motility, which show themselves in cultivation to be facultative anaerobes.

FIG. 98.—ISLETS OF THE BACILLUS OF SWINE ERYSIPELAS ON A GELATINE PLATE. (After Ziegler.)

They take the aniline dyes readily, and prove refractory to decolorising processes, Gram's method included. The latter gives especially fine results with sections. Gelatine is not liquefied, and colonies form on the plate which possess a multiplicity of radiating processes anastomosing with one another, so that they look like bone-corpuscles (fig. 98). Thrust-cultures grow slowly, development commencing beneath the free surface of the gelatine and gradually advancing into the deeper parts, while numerous ramifying fibres proceed out from the thrust-canal, and fill the gelatine in such abundance that it becomes cloudy. Upon agar a delicate layer forms along the inoculated streak. Morbid phenomena can be set up in mice, pigeons, and rabbits by inoculation

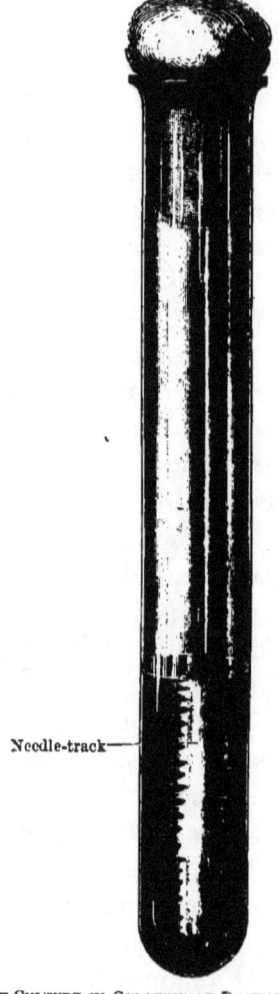

FIG. 99.—THRUST-CULTURE IN GELATINE OF BACILLUS MURISEPTICUS. (After Macé.)

under the skin or into the abdominal cavity, but not in guinea-pigs and poultry.

Bacillus murisepticus.—In putrefying blood and other

BACILLUS MURISEPTICUS

decomposing fluids, Koch found minute immotile rods resembling the bacilli of swine erysipelas, but somewhat thinner and smaller. They look at first sight like fine needle-shaped crystals, and their true nature can only be recognised by staining. They retain their colour when treated by Gram's method. Gelatine is not liquefied, and indistinctly defined colonies occur on the plate and spread out over it in the form of delicate whitish nebulosities. In thrust-cultures in like manner there appear bluish-grey turbid clouds, which permeate the entire gelatine (fig. 99). On agar round isolated yellowish-brown colonies develop

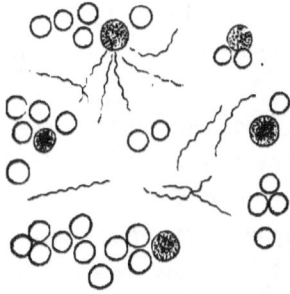

FIG. 100.—SPIRILLA OF RECURRENT FEVER. (After Jaksch.)

along the needle-track. Inoculation causes death in house-mice in two or three days, but field-mice and guinea-pigs are immune. The bacilli are found in the vascular system, and the white corpuscles are often found to be totally destroyed. The blood of the dead animals proves very virulent.

Spirochetæ Obermeieri.—In *recurrent fever* Obermeier found spirilla in the blood at the time of the attack. They are about six times as long as the diameter of a red corpuscle, and move in a very lively manner (fig. 100). They show themselves readily amenable to staining with the ordinary basic aniline dyes, but give up the colour again under Gram's process. A nearly isolated staining may, however, be obtained by employing Günther's method, in which the air-dried cover-glasses are laid for ten seconds in 5 per cent. acetic acid, to bleach the red corpuscles, before being subjected to the action of the staining fluid, after which the acid is removed by blowing, and finally, in order to free the preparation from the last traces of adhering acid, it is laid with the prepared side downwards over an open bottle containing strong solution of ammonia which has just previously been shaken. Staining is then done with gentian violet in aniline water; the preparation is rinsed in water and put up in Canada balsam. Artificial cultivation has not as yet succeeded. Transmission of blood containing spirilla is stated to have produced results only in human beings and monkeys.

Protozoa in the blood.—Protozoa are found in the blood in several infectious diseases. The discovery of Laveran that protozoa have to be dealt with in malarial disorders has been brilliantly confirmed and extended by Celli, Golgi, Saupelice, &c., and is to-day universally admitted. Since malaria is closely connected with the soil, the plasmodia of the disease have already been described in the chapter on the 'Bacteriological Analysis of Earth' (p. 169).

Danilewsky has detected similar forms in the red corpuscles of lizards, tortoises, and birds, and Smith has proved that *Texas fever* in oxen is dependent on a like cause. The *hémoglobinurie bactérienne du bœuf* has, according to Babès, the same significance.

L. Pfeiffer also found protozoa in the blood in *small-pox* and *carcinoma*. [See also Appendix.]

APPENDIX

BY THE TRANSLATOR

A. *Vaccination against Asiatic Cholera*

Principle of anti-cholera vaccination.—The idea of vaccination against cholera is not a new one, as it is now some years since Ferran professed to have discovered a cholera vaccine. Other observers were, however, unable to satisfy themselves of the genuineness of his results, and it is only recently that a process established on a scientific basis has been brought forward by M. Haffkine, of the Pasteur Institute.[1]

During an attack of cholera the specific bacillus is only found in the intestinal tract (see p. 144), and experiments show that it dies when injected subcutaneously. The morbid process must therefore be due to the absorption of toxines generated by it, and Haffkine's vaccination aims at acclimatising the system by the injection of an 'exalted virus,' much stronger than any which it is likely to encounter in the ordinary way of infection, so as to enable it to bear such quantities of cholera poison as may be absorbed from the intestine while an attack of the disease is running its course.

Preparation of the vaccine.—The 'exalted virus' is prepared as follows:[2]—A suspension, in about 3 c.cm. sterile bouillon, of two or more standard cultures (twenty-four hours'

[1] The experiments (on different lines from Haffkine's) of Klemperei, of Vincenzi, and of Brieger and Wassermann should, however, be also mentioned.

[2] Wright and Bruce in *Brit. Med. Journ.* March 4, 1893, p. 227.

growth at 35° C. on an agar surface 10 cm. long) of cholera bacillus having been drawn into a small pipette, the point of the latter is bored through the abdominal wall of a guinea-pig, at a spot previously sterilised by cauterisation after clipping the hair short, and the contents are blown out into the peritoneal cavity. Immediately after death (which takes place in twenty-four hours) the abdomen is opened with the strictest antiseptic precautions, and the fluid found in the iliac fossæ is transferred by means of a sterile pipette to a sterilised test-tube, in which it is left in the incubator for 8 or 10 hours at 35° C., in such a position that the largest possible surface of fluid is exposed to the air. The process of intra-peritoneal injection, &c., is then repeated, and this is done twenty to thirty times in succession, until the highest possible degree of virulence is attained, which may be over a hundred times as great as that of the original growth. Cultures of the exalted virus can then be made, but do not retain their full virulence for many generations.

The vaccine actually injected may contain the living bacilli (since these perish in the tissues), or may be 'carbolised,' so as to kill them. Carbolised vaccines are less dangerous and liable to contamination, and may be kept indefinitely in sealed tubes, but are not so powerful. The vaccine is thus obtained:—A standard culture is made by inoculating the whole surface of an agar tube and keeping it at 35° C. for 24 hours, after which an emulsion is prepared by mixing the growth with 2 or 3 c.cm. of bouillon, and then diluting up to 8 c.cm. for living vaccines; or with 6 c.cm. of sterile ½ per cent. carbolic acid solution for carbolised vaccines. The amount of either to be injected is 1 c.cm., and the injection should be made under the skin of the flank.

Results of vaccination.—Injection of such a vaccine causes a rise of temperature accompanied by malaise and other slight general symptoms, which soon pass off; locally, however, severe inflammation and necrosis follow, unless a preliminary injection has been made 3 to 5 days previously with a weak vaccine prepared from cultures attenuated by being grown in media kept continually aërated, and at 35° C. The only local symptoms are then slight pain and œdema. After the symptoms have passed off the animal is found to be scarcely affected by many times larger intraperitoneal injections than suffice to kill control animals not so protected.

These results have also been obtained with rabbits and pigeons, and the symptoms following injection into human beings, of whom about a hundred had been vaccinated up to March last,[1] are identical with those exhibited by all the animals, subsequent hypodermic injections with the exalted virus also producing the same results in both, so that although, of course, human beings could not be directly tested like the animals, there seems little room for doubt that they are similarly protected. It may be added that swallowing a draught of cholera bacilli in one case, and free intentional exposure to contagion at Hamburg in another, produced no results in two observers who had been vaccinated.

More recently, however, some doubt has been cast on the efficacy of the process to protect against the ordinary infection of cholera by Klein,[2] who found that precisely similar results could be obtained with *Vibrio proteus*, the *typhoid bacillus*, *Bacillus coli*, *Bacillus prodigiosus*, &c. Vaccination with an exalted virus prepared from any of these conferred immunity against the cholera bacillus (even the exalted virus) or any of the others, when injected intraperitoneally. He also found that guinea-pigs protected by Haffkine's method were killed by intraperitoneal injection of a

---

[1] Haffkine, *Fortnightly Review*, March 1893.
[2] *Brit. Med. Journ.*, March 25, 1893, p. 632.

gelatine culture of the cholera bacillus liquefied by its growth. From these observations he concludes that Haffkine's vaccination only affords protection from the intracellular poison, which is the same in all the bacteria tried, but not from the poison generated externally, which differs in all the varieties, and is the cause of the distinctive disease.

It was, however, pointed out in reply [1] that the products of gelatine cultures would be more or less special, and that Haffkine's method protects against cholera virus introduced into the alimentary canal; while the conclusion regarding the identity of the intracellular poisons in all the bacilli named is not regarded as unimpeachable.

B. *Parasitic Protozoa*

Pathogenesis of protozoa.—We have already seen that a parasitic protozoon may originate morbid processes, in the case of the *Plasmodium malariæ* (pp. 169 and 282). There are, however, a number of other protozoa which frequent the bodies of men and animals, and some of these are believed, with greater or less degrees of probability, to be the cause of certain diseases. The best known of these is the *Coccidium oviforme*, which is important as throwing light upon the habits of other less known protozoa, and more especially as showing that the presence of such an organism may cause proliferation of epithelial cells.

Coccidium oviforme.—The life-history of this parasite, which causes a very fatal disease in young rabbits, has quite recently been worked out by R. Pfeiffer,[2] although its existence was known as far back as 1839.

When discharged from the affected rabbit in the fæces, it is a firm, translucent, oval cyst measuring about $36 \times 22\mu$, with granular contents. These contract into a ball, and by

---

[1] *Brit. Med. Journ.*, March 25, 1893, p. 639.
[2] *Beiträge zur Protozoen-Forschung*, Heft i. Berlin, 1892.

pushing out projections divide into four parts, each of which subdivides into two crescentic germs and a nucleus, and becomes surrounded by an inner capsule. In this condition the coccidium may remain unaltered for months unless swallowed, in which case the germs are set free by the digestion of the capsules, become rounded and amœboid, and usually divide into crescentic germs, which also become free; the result being that the intestine, gall-bladder, and bile-ducts are filled with sporozoa. The sporules finally penetrate into the epithelial cells of the mucous membrane, and there change back into the encysted form, which is set free by the bursting of the cell.

During their growth they produce great proliferation of the neighbouring epithelium, causing the formation in the bile-ducts of the liver of greyish-white adenomata, which consist of epithelium and connective-tissue and are bounded by a connective-tissue layer, the interstices between the branching processes which grow from the mucous coat being full of parasites in various stages of development. A hypertrophy of the mucous membrane also takes place over the affected areas of the small intestine. If the animal recovers, the parasites are discharged or absorbed, being in the latter case penetrated and digested by small-cell infiltration,[1] and a scar of connective tissue is left on the site of the tumour.

Feeding with the ripe coccidia is the only method of infection which has been attended with success. The changes in the parasites may be observed by taking some from recently killed animals, placing on a cover-glass, and examining in drop cultivations.[2]

Amœba dysenteriæ.—The occasional occurrence of proto-

[1] Ruffer and Walker, *Journ. of Pathol. and Bacteriol.*, October 1892.

[2] The above account is mainly taken from the Morton lecture of this year, by Dr. Galloway. See *Brit. Med. Journ.* February 4, 1893, p. 217, for bibliography, &c.

zoa in the stools is a fact that has long been known, but Lösch was the first to connect them with *dysentery*, in a paper published in 1875 and describing a case of that disease in which they were present. Many observations have been since made, the most important communications on the subject being those of Kartulis from Egypt, and of Councilman and Lafleur in America (to the latter of which the above name is due, Lösch having called the protozoon *Amœba coli*). The variety of the disease ascribed to its agency is that known as *tropical dysentery*, which differs both clinically and pathologically from other forms, and the amœba was found by Kartulis [1] in 500 cases of this prior to the appearance of his second paper, as well as in every case of dysenteric liver abscess examined, while it was absent from 'idiopathic' liver abscesses; and Councilman and Lafleur [2] found it in all the fourteen cases on which their very elaborate monograph is based.

According to the latter observers, the amœbæ when at rest are round or slightly oblong bodies, consisting of an outer pale homogeneous substance enclosing a somewhat greenish highly-refractive mass, which contains vacuoles of various sizes and a nucleus. Movement is their distinctive feature, however, and consists first of a progressive motion, and secondly of a protrusion and withdrawal of pseudopodia, both of which vary in activity. The pseudopodia are formed from the outer homogeneous part, which may, however, be otherwise invisible both in the resting and moving state. The amœbæ often contain foreign bodies, such as red corpuscles, pus-cells, blood-pigment, micrococci, bacilli and their spores, &c.

Entering probably with the food, they pass on until the large intestine is reached, where the alkalinity necessary

[1] *Virchow's Archiv*, Bd. 118, 1889.
[2] *Johns Hopkins Hosp. Reports*, vol. ii. Nos. 7, 8, 9.

to their growth is obtained. Here they penetrate and undermine the mucous membrane, producing their effects by liquefying the tissues, and thus causing ulceration and necrosis. In the mucous membrane they are found chiefly in the submucosa, in the lymph-spaces and blood-vessels, and in the gelatinous contents of the ulcers. They may penetrate to the liver either by the portal vessels or through the peritoneum (sometimes causing peritonitis), and set up abscesses, which in the former case may be multiple, in the latter lie close to the surface of the right lobe, the commonest position. According to Councilman and Lafleur, the liver shows no inflammatory reaction, and the abscess-cavity is filled, not with pus, but with *débris* of liver tissue, or sometimes a necrosed mass; the contents may, however, be old pus, the suppuration being due to the action of micro-organisms conveyed by the amœbæ (Kartulis). The liver abscess may extend directly so as to involve the lung, or the amœbæ may traverse the diaphragm and set up an abscess by liquefying the tissue; but here the cavity is surrounded by an area of interstitial inflammation. That the micro-organisms do not travel by the lymphatics is shown by their absence from the mesenteric glands; and they have no special preference for the lymphoid follicles.

Ante-mortem they may be found in the stools, particularly the gelatinous particles, in the pus from liver abscesses, and in the sputum in cases of abscess of the lung, and may be examined in the fresh state, best on the warm stage, or in cover-glass preparations; or sections may be cut from portions of fæces hardened and imbedded in celloidine. Councilman and Lafleur obtained the best results with sections of tissue hardened in alcohol and stained in Löffler's methyl blue, by which method the amœbæ are coloured dark blue, but unevenly. The nuclei are best brought out by hardening in Flemming's solution (see

Section D), and staining deeply with saffranine, and a nucleolus can often be seen.

Kartulis succeeded in obtaining pure cultures by inoculating alkaline infusion of straw with some of the contents of a liver abscess in which no bacteria were present. In twenty-four to forty-eight hours, at 35° to 38° C., a membrane forms on the surface, consisting of young amœbæ, which are mixed with bacteria when the culture is not quite pure. Injection of a pure culture into the rectum, followed by tying of the anus, caused swelling and erosions of the mucous membrane in cats. Rectal injections of dysenteric stools have also been often, though not always, successful in causing similar diseased conditions. No result followed when amœbæ were administered by the mouth.

Protozoa in carcinoma.—As far back as 1847 Virchow observed certain inclusions in tumour cells, some of which are judged from his plates to have been similar to the bodies now regarded in many quarters as parasitic protozoa. They were variously explained at the time, and it was not until 1888–89 that their nature began to be seriously discussed in connection with a possible causative influence in cancer. A number of observations were published in 1889, perhaps the most important being that of Malassez and Albarran (in the affirmative), and since then many researches have been carried on, amongst the more recent of which the observations of Ruffer and Walker[1] in England especially (which have been since confirmed by Burchardt and Foà), and those of Soudakewitch and of Steinhaus on the Continent, must be mentioned. Many of the observers seem to have been misled by various other appearances, but some of the earlier investigations and most of the later concur in showing that a peculiar body is practically always present

---

[1] *Journ. of Pathol. and Bactcriol.*, October 1892. This paper is beautifully illustrated with chromo-lithographs, which see.

in some of the cells of all varieties of carcinoma, although the number is not constant.

These inclusions are found for the most part in the body of the cell, usually singly, though as many as eight or ten have been counted, and they appear as round or oval figures, measuring from 2μ to 10μ or more in diameter, surrounded by a capsule probably furnished by the epithelial cell (Ruffer and Walker), and having a large irregular nucleus in the centre, from which processes extend out radially. Lines are also visible, at regular intervals, running in from the capsule towards the centre. Less frequently the inclusions are seen in the cell nucleus, or sometimes half in and half out, a condition in which the capsule is slight or absent (Galloway),[1] and sometimes the nucleus contains a number of small ones, which eventually make their way out. Still more rarely similar bodies are observed in the intercellular spaces. The nuclei of the bodies stain like the nucleoli of ordinary cells when treated by the Ehrlich-Biondi method (see below).

Ruffer states that reproduction takes place by repeated division of the nucleus, followed by that of the capsule, until a zoogloea-like mass of 'parasites' is produced. He has not observed formation of spores, but thinks that some of the young 'parasites' may be of this character. Burchardt,[2] however, on one occasion found an appearance in one of the bodies (which he has figured), consisting of a delicate oval outline, which contains a round, thick-walled vesicle full of small round bodies, and these he believes to be a spore, germ-capsule, and germs respectively.[3]

The nucleus of the cancer cell is pressed to one side, as

[1] Morton Lecture, *Brit. Med. Journ.*, February 4, 1893, p. 217
[2] *Virchow's Archiv*, Band 131, p. 121, January 2, 1893.
[3] Foà has described the division of the body into a number of highly-refracting particles *within the capsule*, which he regards as spores, and which eventually penetrate into fresh cancer cells.

a rule, but the cell may otherwise remain normal, or it may become vacuolated and finally converted into a cyst. The cancer cells in the neighbourhood do not show greater signs of activity than those elsewhere (Boyce), but on the other hand the parasites are most numerous in rapidly-growing cancers, in newly-started secondary tumours, and at the growing edges and outlying parts.

The 'parasite' itself does not seem to be very hardy, at times appearing eaten away; and Ruffer and Walker have also seen it attacked and destroyed by leucocytes, which sometimes surround it and sometimes penetrate into its interior.

The body, therefore, is treated as a foreign substance in the tissues, looks like an organised structure, competent zoologists (Balbiani and Metschnikoff) have pronounced for its parasitic nature, and appearances resembling germination have been observed. It also stains unlike normal cell-contents, resembling in this, as well as in some other respects, known sporozoa. On the other hand, it differs both in size, durability, and some other particulars from the protozoa with which it can best be compared, has never been seen in an amœboid stage, and its sporulation is scarcely fully established as yet. Finally, cultivation has not yet succeeded, although certain bodies of a doubtful nature have been observed by Ballance and Shattock [1] to undergo multiplication in a piece of scirrhus of the breast kept at incubation temperature. The opponents of the parasitic theory refer it to one of the following classes, with which it certainly seems to have at times been confounded: [2] transverse sections through two cells, one of which is invaginated into the other; leucocytes, or red corpuscles, enclosed in

[1] Report of debate at London Pathological Society, *Brit. Med. Journ.*, March 11, 1893, p. 520.
[2] See Galloway, *loc. cit.*

epithelial cells; endogenous cell-formations, in which the nucleus has divided, but not the cell; asymmetrical mitoses; degenerations of the cancer cells, particularly if the nucleus is affected. Upon the whole, it may be said that the weight of probability is on the side of the parasitic view, but its influence on the tissues is at present unknown.[1]

The pieces of tissue to be examined should, if possible, be obtained fresh from the operating table, cut into very small pieces, and immersed at once in absolute alcohol, corrosive sublimate, or Flemming's solution (see Section D). If either of the latter be used, hardening must be completed in alcohol. The pieces may then be imbedded in paraffine or celloidine, and sections cut and stained.

Hæmatoxyline stain alone, or better with eosine or rose bengale double-staining, gives good results, especially after hardening in Flemming's solution. The parasites show a different tone of blue from the cell-nuclei.

By Russell's method the sections are stained for ten minutes in saturated solution of fuchsine in 20 per cent. carbolic acid, washed for a few minutes in water and then for half a minute in absolute alcohol, and transferred for five minutes to a 1 per cent. solution of iodine green (Grübler's) in 2 per cent. carbolic acid. The sections are then rapidly dehydrated in absolute alcohol, cleared in oil of cloves, and mounted in balsam. The 'parasites' appear purple or red, the tissue-cells blue; but unfortunately some other bodies also stain red and may lead to mistakes.

The Ehrlich-Biondi triple stain, which gave beautiful results in the hands of Ruffer and Walker, had better be bought prepared by Grübler of Leipsic, as it is rather hard to make properly. The following was the method employed by the observers just mentioned:—1 grm. of

---

[1] For review of the general arguments in favour of a parasitic origin of cancer, see *Ann. of Surg.*, vol. xvii. pt. 4 (April 1893), p. 478.

Grübler's powder is dissolved in 80 c.cm. of water, and 15 c.cm. of a 0·5 per cent. solution of acid fuchsine is added. The sections are stained for at least an hour, washed in water for half a minute, placed in 95 per cent. alcohol for one minute and then for two to five minutes in absolute alcohol, and finally transferred to xylol and xylol balsam.

Sims Woodhead gives the following method of preparing and using this stain:—5 c.cm. saturated solution of methyl green, 10 c.cm. saturated solution of methyl orange, and 2 c.cm. saturated solution of acid fuchsine are mixed, after having first been diluted with about 40 volumes of water each, to avoid the formation of a precipitate. Stain for fifteen minutes to twelve hours, rinse in 1 per 1,000 acetic acid, wash in dilute and then for one minute in absolute alcohol, and finally immerse in xylol and benzol, and mount in xylol balsam.

The protoplasm of the epithelial cell is orange-red, the nucleus green, the nucleoli brown or red; the protoplasm of the 'parasite' is pale blue and the nucleus red. Connective tissue is red, and leucocytes and all other cells are orange-red with green nuclei. The protoplasm of the 'parasite' is, however, sometimes orange-red, and the nucleus a little darker. Lastly, the living parasites may sometimes be distinctly seen, according to Soudakewitch, by examining scraped-off cancer cells in 0·6 per cent. salt solution.

(For bibliography, refer to the papers by Ruffer and Walker, and by Galloway.)

Protozoa in other forms of new growth.—Whether bodies similar to those just described occur in *sarcoma* is still a matter of doubt. Jackson Clarke, indeed, claims to have found such, but the Morbid Growths Committee of the London Pathological Society has decided [1] that his supposed

---

[1] See *Report, Brit. Med. Journ.*, May 20, 1893, p. 1056.

APPENDIX 295

parasites are really giant cells which have undergone necrotic changes.

Peculiar bright bodies which are visible in the masses of *molluscum contagiosum*, lying partly free and partly in the interior of horny cells, are held by many to be coccidia, and to be the exciting cause of the diseased growth. Others, however, believe them to be merely degenerated epithelial elements.

Bodies of a similar nature have also been described (by Darier, Wickham, and others) in the form of eczema of the nipple known as *Paget's disease*, which is associated with cancer of the mammary ducts, but Ruffer and Walker [1] are of opinion that many at least of the appearances are degenerating epithelial cells, or due to endogenous cell formation.

Wernicke has described coccidia in the granulomatous tumours of *mycosis fungoides*. These are light yellow, round, 3 to 30 μ in diameter, and are contained in a hyaline sheath. As many as ten have been observed in a giant cell. They reproduce by segmentation, and the young coccidia are set free by the bursting of the parent cells.[2]

C. *The Action of Light upon Micro-organisms*

The action of white light.—The fact that certain effects are produced upon micro-organisms by the sun's rays has been known for some time; thus certain pathogenic bacteria show less marked activity after exposure to light, although on the other hand some bacteria, probably only the aquatic varieties (Buchner), grow better under these circumstances; and others belonging to the chromogenic species—*Bacillus prodigiosus*, for example—cannot produce their pigment in

[1] *Loc. cit.* [2] *Centralb. f. Bakt. u. Parasitenk.*, Dec. 28, 1892.

the dark. The first definite researches on the subject, however, were made in this country by Downes and Blunt in 1877, since when many investigations have been carried on, with the result that the power of light to inhibit the growth of many micro-organisms has now been placed beyond a doubt.

Thus Koch [1] found that *tubercle bacilli* are killed on exposure to direct sunlight, varying from some minutes to a few hours, according to the thickness of the layer, while diffuse daylight required from 5 to 7 days to produce the same effect. Similar results have been obtained with the *typhoid bacillus* by Janowski,[2] who exposed to diffuse daylight three culture tubes, one uncovered, one wrapped in white, and one in black paper, and also a culture in a U tube one limb of which was covered, and found that the uncovered culture or portion of a culture was greatly retarded in its development, while 4 to 7 hours of direct sunlight killed it altogether. Buchner[2] exposed a plate culture of the same bacillus in a Petri's capsule, to which a cross of black paper had been attached, and then kept it for 24 hours in the dark, when it was found that after 1 to 1½ hour of sunlight, or 5 of diffuse daylight, no colonies developed except on the part of the plate protected by the paper cross, all the rest having been destroyed. He also obtained confirmatory results with *Bacillus coli commune*, *Bacillus pyocyaneus*, and *Vibrio cholerae Asiaticae*, finding that some water which contained 100,000 elements of the first-named per c.cm. yielded no growth after an hour in

[1] *Verhandlung. d.* 10. *medic. Congress. Berlin*, 1890, Bd. I. p. 42 (quoted by Günther).

[2] For bibliography up to date, see P. Frankland and H. M. Ward, 'Report on the Bacteriology of Water,' *Roy. Soc. Proc.*, vol. li. No. 310, pp. 199 *n*. and 237–9; also the other papers by the latter (to be cited presently). It may be mentioned, however, that Janowski's paper appeared in *Centralb. f. Bakt. u. Parasitenk.*, Nos. 6–8, 1890; and Buchner's *ibid.* Bd. xii. 1892, and Bd. xi. No. 25.

direct sunlight. The most recent and conclusive experiments of all, however, are those of Professor H. Marshall Ward [1] with the very resistent and virulent spores of *Bacillus anthracis*, confirming and extending the results obtained by some previous observers with the same bacterium. Having found that repeated exposure to sunlight destroyed the spores in a few c.cm. of Thames water containing a very large number, while a few weeks of bright daylight greatly lessened them, he proceeded to make a series of accurate experiments as follows:—Agar plates of anthrax were made in Petri's capsules, using the virulent and resistent spores obtained by keeping some c.cm. of distilled water, saturated with material from an old agar culture, at 56° C. for 24 hours. To the bottom of each plate was attached a zinc stencil having a letter cut out, the rest of the capsule was covered with dull black paper, and the letter surface exposed to direct or reflected sunlight for 2 to 6 hours, after which the plate was kept in the incubator at 20° C. for 48 hours. The agar was then found to be grey and cloudy from the development of a multitude of colonies, but the space exposed to light remained quite clear, showing the form of the letter, and the development of the colonies in its neighbourhood were greatly retarded, owing to the action of reflected light, when the letter was large. These results were also obtained with other bacteria, as well as with the spores of fungi; and similar, though less marked, effects were produced by the electric arc light, so much so that Marshall Ward thinks that it may yet prove to be an efficient disinfecting agent.

The action of coloured light.—When a plate culture of anthrax spores is exposed to the solar spectrum the bactericidal action is found to be strongest at the blue-violet end (Ward), and certain chromogenic bacteria have been

[1] *Royal Soc. Proc.*, vol. lii. No. 318, p. 393.

seen,[1] when examined in water by the micro-spectral objective, to make their way to the part furthest from this end. Janowski exposed cultures under screens of various chemical solutions and aniline dyes, and found that no action took place under brown or yellow, whereas solutions of fuchsine (which transmits violet rays), gentian violet, and methyl blue had little more effect than colourless fluids. Confirmatory results have also been obtained quite recently by Marshall Ward,[2] who repeated his experiments with the anthrax plates, covering the stencil letter with pieces of glass of various colours, as well as with screens of saturated solutions of potassium bichromate and of ammoniated cupric oxide (which cut off respectively the violet and red of the spectrum as far as the line b), and found that no effect was produced on the spores by exposure to sunlight behind ruby, olive, orange, or green glass, nor behind the bichromate solution (when sufficiently concentrated), while 4 hours' exposure behind violet and blue glass, and a much thicker cupric oxide screen, was usually sufficient to kill them. Less pronounced effects were produced by diffuse daylight. The bactericidal action of light, therefore, seems to depend on the more refrangible rays in the violet half of the spectrum, which produce their effects whether the red-yellow rays are transmitted or not.[3]

The mode of action of light.—That the effects produced are really due to the rays of light and not to heat is shown by the fact that they also take place under water (Buchner) and behind a screen of alum solution (Janowski), that a thermometer at the surface of the glass never registered above

[1] Engelmann, quoted without reference by Woodhead (*Bacteria and their Products*, 1892, p. 201).
[2] *Roy Soc. Proc.*, vol. liii. No. 321, p. 23.
[3] Prof. Ward states (May 25) that he has made great advances in his experiments with chemical screens and with the spectrum, but has elicited no contradictions to his earlier results, as already published.

18° C., and that a gelatine plate with a melting-point of 29° C. remained solid during 5 or 6 hours' exposure, while the spores in the exposed area were killed (Ward). An alternative, explanation, that the death of the spores may be due to changes brought about in the nutrient medium, has also been excluded by Marshall Ward, who proved that no colonies develop (provided insolation has been complete) if a piece of the clear letter on the agar is placed in fresh nutrient material, while it is still capable of supporting a growth of the bacillus if sown with fresh spores. He also exposed a plate of dry spores and one of sterile agar under stencil letters side by side, and then covered them respectively with gelatine or agar, and with fresh spores, when the former showed a clear area under the letter, whereas the latter was covered by a uniform growth. The effects are therefore due to the direct action of the rays of light upon the micro-organisms themselves.

It is probable that ripe spores contain a reserve store of fatty material, which would partly explain their resistance to staining and to the ordinary methods of sterilisation, and Duclaux has shown that vegetable oils are rapidly oxidised by light. From these considerations, together with the facts that the effect is stronger on spores than on bacteria and has the character of chemical action, Ward has deduced the theory that the bactericidal action of light may be due to its destructive influence, in the presence of oxygen, on fatty matters or other reserve material.

Applications in Nature.—The above theory seems to explain and be supported by many facts in nature. Thus chromogenic bacteria often produce their pigment only in the presence of light, that is, when it would be required to act as a screen, and fungi which grow in open places and on the upper surface of leaves are nearly always protected by having a dark, or a yellow-red, or an orange colour. Ward could not obtain any positive results with fungus spores which were of a deep colour, or dull yellow-brown, although colourless and faintly tinged spores were readily killed ; and Bachmann found that in the spectra of 42 pigments of fungi the extreme blue-violet (i.e. the most active part) is cut out from the line G on in all but two.

Further, the sporangia and spores of ferns are mostly orange or reddish, pollen and its spores, which contain oils, are almost always deep orange, and lastly Ward suggests that one of the functions of chlorophyll may be to act as a colour-screen.

This observer therefore puts forth the provisional hypothesis that :—' No plant exposes a reserve store of fatty material to the danger of prolonged or intense insolation without a protective colour screen calculated to cut out at least the blue-violet rays, as these rays would otherwise destroy the reserve substance by promoting its rapid oxidation.'

D. *Additional Methods and Formulæ*

Fixing methods.—For fixing coccidia, Flemming's solution, corrosive sublimate, or absolute alcohol may be used. The last-named was dealt with on p. 80.

By Flemming's method pieces of perfectly fresh tissue are immersed in a mixture of 4 parts of 2 per cent. aqueous solution of osmic acid, 15 parts of 1 per cent. chromic acid, and 1 part of glacial acetic acid, for from one to three days, washed in water for three to six hours, and then finally hardened by successive immersion for twenty-four hours each in 30, 60, and 96 per cent. alcohol. This method gives great sharpness of definition, but the sections do not stain readily.

By the second method the fresh pieces are immersed for six to twenty-four hours in a solution of $7\frac{1}{2}$ grms. corrosive sublimate in 100 parts of 0·7 per cent. salt solution, washed first in water, then in 30 per cent. alcohol to which a few drops of tincture of iodine have been added, and then in weak potassium iodide solution, and hardened by immersion in 50, 75, and 90 per cent. alcohol for twenty-four hours each, and finally in absolute alcohol. Imbedding may then be done in paraffin or celloidine.

The gum freezing method.—A very convenient method of cutting sections, which is much used in this country, is that of freezing in gum. The piece of tissue, which should be as small as possible, is soaked for twenty-four hours in water to free it from spirit. It is then left in thick gum mucilage until completely saturated, which requires from a few hours to some days, according to the density of the tissue, after which it is frozen on the plate of the microtome, and in that state has exactly the right consistence for cutting. The sections are transferred to tepid water to free them from gum, and can then be stained, &c., as usual, or kept in spirit. Fresh or imperfectly hardened tissues can be cut with ease by this method.

Staining formulæ.—Von Kahlden recommends the following method of staining sections of tissue fixed in Flemming's solution in order to bring out the nuclei:—The sections are stained for half an hour to twenty-four hours in 1 per cent. aqueous solution of saffranine, rinsed for a short time in water, washed in absolute alcohol slightly acidulated with a few drops (5 to 10 drops to a medium watch-glass) of spirit containing 1 per cent. of hydrochloric acid, and finally in absolute alcohol, cleared in xylol, and mounted in Canada balsam.

The simplest *hæmatoxyline* stain is made by adding sufficient of a saturated alcoholic solution of the crystals to a 1 per cent. aqueous solution of alum to render it pale blue or pale violet, and exposing to light for a few days. Filter, and stain for two to three minutes. Ehrlich's *acid hæma-toxyline*, which has been much used for staining coccidia, is prepared by adding 6 parts of glycerine and 60 of water, both saturated with alum, and 3 parts of glacial acetic acid, to a solution of 2 parts hæmatoxyline in 60 parts absolute alcohol. Filter, stain for 3 to 5 minutes.

Eosine is used as a double stain in a $\frac{1}{10}$ per cent. aqueous solution in which the sections, after being stained in hæmatoxyline and washed, are immersed for one or two minutes.

Picro-carmine is made by dissolving 1 grm. carmine in 5 grms. strong ammonia and 50 grms. water, and adding 50 grms. of saturated aqueous solution of picric acid, after which the stain is left standing in a wide open vessel until all the ammonia has evaporated, and is then filtered. Stain for an hour, soak for half an hour in glycerine containing 1 per cent. of hydrochloric acid and coloured light yellow with picric acid, wash for five minutes in water, and dehydrate in alcohol, both of which are similarly tinted.

Alum carmine solution is made by boiling 2 to 5 grms. carmine with 100 grms. 5 per cent. alum solution for a half to one hour, and filtering. Stain for ten minutes to two hours.

Nicolle claims to have discovered a method of fixing the colour of bacteria, useful for those which bleach under Gram's process:—The sections are stained for two or three minutes in Löffler's or Kühne's methyl blue, slightly decolorised in feebly acid water, washed, dipped for an instant in 10 per cent. tannin solution, washed in water, dehydrated, cleared in oil of cloves, and mounted in xylol Canada balsam.

INDEX

ABSCESS, hepatic, in dysentery, 289
— pulmonary, 289
Abscesses in skin, staining micro-organisms in, 90
Aceti, Bac., 191
Achorion actaton, dicroon, euthythrix, Schœnleinii, 234
Acidi lactici, Bac., 179 ; *Mic.*, 180
— — *liquefaciens, Mic.*, 181
Actinobacter, Bac., 185
Actinomyces, 196, 275
Aerobes, 1
Acrogenes, Mic., 250 ; *Bacter., Helicobacter, Bac.*, 251
Aëroscope, Pouchet's, 97
Agar, glycerine, 45 ; Kowalski's, 46 ; litmus, 45 ; meat extract peptone, 45 ; peptone bouillon, 43 ; serum, 49 ; surface cultures on, 57 ; urine, 46
Agar-agar, 42
Agilis, Mic., 128
Air, analysis of, 96 ; methods of examining, 96 ; method of Pouchet, 97 ; of Miquel, 98 ; Emmerich, Welz, 99 ; Hesse, 100 ; Strauss and Würz, 101 ; Petri, Tyndall, 102 ; Frankland, 103 ; micro-organisms in, 96
Alba, Sarcina, 108
Albicans amplus, Diploc., 225
— *tardus, Diploc.*, 226
Albumen, plover's egg, 49
— solutions of, 37
Albuminate, Lieberkuhn's potash, 49
Albuminis, Bac., 256
Ambigenous organisms, 1
Amœba coli, v. dysenteriæ, 252, 288
Amylobacter, Bac., 181
Anaerobes, 2 ; cultivation of, 60

Aniline colours, 30 ; A. oil, A. water, 69
Animals, experiments on, 91 ; most useful, 92
Anthracis, Bac., 165 ; effect of light on, 297
Anthrax, symptomatic, Bac. of, 228
Apparatus, list of, 20 ; counting, 125 ; plate, 28, 55 ; spray, 93
Aquatilis, Bac., 136 ; *Mic.*, 128
— *radiatus, Bac.*, 137
— *sulcatus, Bac.*, 136
Arborescens, Bac., 131
Arthrospores, 5
Ascobacillus citreus, 231
Aspergillineæ, 10
Aspergillus albus, flavescens, fumigatus, glaucus, niger, 193
Attenuated vaccine, Haffkine's, 285
Attenuation, 6
Aurantiaca, Sarc., 108
Aurantiacus, Bac., 135
— *Mic.*, 129
Aurea, Sarc., 270
Aureus, Bac., 135, 275

BACILLI, 2
Bacteria, varieties of, 2 ; multiplication of, 4
Bacteriuria, 258
Basidia, 11, 103
BAUMEYER, petroleum incubator, 26
BAUMGARTEN, staining of tubercle bacilli, 215
Beggiatoa, 12
Birds, tuberculosis in, 209
Biskra, Mic., 221
Blood, micro-organisms in, 276 ; modes of examining, 276 ; staining of, 277
— serum. *See* Serum

Borax methyl blue, method, 90
Bougies, œsophageal, 94
Bouillon, 35; meat, 36; meat extract, 37
Bread, 53; moulds on, 193
BRIEGER, extraction of toxines, 7
'Brownian movement,' 3
Bruneus, Bac., 135
Buccalis maximus, Bac., 238
Butyri fluorescens, Bac., 183
— *viscosus, Bac.*, 183
Butyricus, Bac., 182

CADAVERINE, 148
Candicans, Mic., 106
Candida, sarc., 108
Capsulatus, Bac., 3, 221
— *mucosus, Bac.*, 3, 268
Capsules, 3
Carcinoma, protozoa in, 282, 290
Carmine, alum, 302
Carneus, Mic., 129
Carragheen moss, nutrient jelly from, 47
Caucasicus, Bac., 188
Cavicida, Bac., 256
Celloidine, imbedding in, 82
Cells, inclusions in, 290
— epithelial, proliferation of, caused by coccidia, 287
Centrifuges. *See* Machines, centrifugal
Cereus, Staph., albus, and *flavus*, 199; *aureus*, 201, 264
Cerevisiæ, Sarc., 192
'Chamberland's candles,' 123
'Chameleon phenomenon,' 199
CHENZYNSKY's solution, 86, 278
Cholera, inoculation of, 94, 144; vaccination against, 283; reaction, 144, 147
Cholera Nostras, Bac. of, 144, 145
Cholera-red, 147
Choleræ Asiaticæ, Bac., Vibrio, or Spirillum, 139; effect of light on, 296; toxopeptone from, 7
Choline, 148
Cilia. *See* Flagella
Cinabareus, Mic., 106
Citreus conglomeratus, Dipl., 107
— *liquefaciens, Dipl.*, 226
Cladothrix, 12; *C. dichotoma*, 13; *C. canis*, 197
Clearing agents, 85
Clostridium, 4, 159; *C. butyricum*, 181, 254; *C. fœtidum*, 159
Cocci, 2

Coccidia, 12, 295
Coccidium oviforme, 286
Coli commune, Bac., 152; effect of light on, 296
Colonies on gelatine plate, 56
Coloured images, 68
— light, action of, on micro-organisms, 297
Columella, 10
Comma-bacilli, 2, 140, 146
Concentricum, Spir., 176
Concentricus, Mic., 129
Condensation, water of, 45
Conidia, 11, 103
Conjugation, reproduction by, 194
Coprogenes fœtidus, Bac., 251
Coryzæ, Dipl., 265
— *equorum contagiosæ, Mic.*, 221
Counting apparatus, Wolffhügel's, 125
Crates, wire, for test-tubes, 29
Crenothrix Kühniana, 11
Cultivation, methods of, 53–64
Cultures, anaerobic, 60; 'hanging drop,' 62; high, 62; permanent, 63; plate, Koch's, 54; plate, serum and albumen, 60; roll, Esmarch's, 58; slide, 53; streak or surface, 57, 60; thrust, 57, 62
Cumulatus tenuis, Mic., 264
Cyanogenus, Bac., 185
Cystitis, Bac. in, 262

DECOLORISING agents, 72
Decomposition, 173
Dehydration, 85
DENEKE, Bac., 146, 183
Denticola, 242
Dentium, Spirochæte, 242
Diet, articles of, 178, 189
Digestive tract, the, 237; infection by, 93
Diphtheriæ, Bac., 240
Diplococci, 2, 4
Discontinuous sterilisation, 18
Disinfectants, chemical, 18
Dispora Caucasica, 188
Double test-glasses, Schill's, 16
Drum-stick bacteria, 5
Drying-on processes, Unna's, 85; Kühne's, Weigert's, 89; Winkler's, 203
Dust, window and room, 156
Dysenteriæ, Amœba, 252, 288
— *Bac.*, 252

EARTH, analysis of, 155
— bacillus, 156

INDEX 305

EBERTH, bacillus of typhoid fever, 148
Ectogenous organisms, 1
Ectoplasm, 171
Eczema marginatum, 233
Eggs, birds', hens', 49, 50 ; plover's, 49
EHRLICH-BIONDI triple stain, 293
EHRLICH, acid hæmatoxyline, 302
Emmerich, Bac., 120
EMMERICH, method of air-analysis, 99
Endocarditis capsulatus, Bac., 278
Endogenous organisms, 1
Enteric fever, 148. *See* Typhoid
Entoplasm, 171
Eosine, double-stain, 302
ERLENMEYER'S flasks, 35
Erysipelatis, Strept., 113
Erythrosporus, B., 131
Eye, infection into anterior chamber of, 95

FACULTATIVE anaerobes, 2
— parasites, 1
Fæces, composition of, 254 ; protozoa in, 255
Favus, 234 ; fungus of, 234, 235
Febris tertiana, quartana, quotidiana, 171
Fermentation, 12 ; acetic, 191 ; alcoholic, 12, 188 ; alkaline, 258 ; butyric, 182; lactic acid, 180, 188, 192 ; slimy, 184
Fervidosus, Mic., 129
Figurans, Bac., 175
Filters, hot-water, 27 ; creased paper, 38 ; sand, charcoal, asbestos, porcelain, clay, 122 ; kaolin (Chamberland-Pasteur), 123 ; micro-membrane (Breier), 123
Filtration, 122 ; of media, 38, 43
Finkler-Prior, Bac., 145
'Fishing,' 57
Fission-fungi, 2, 4. *See* Bacteria
Fission, multiplication by, 4
Fixing tissues, methods of, 300 ; serial preparations, media for, 84
Flagella, 3 ; staining of, 71
Flavus desidens, Mic., 108
— *liquefaciens, Mic.*, 108
— — *tardus, Dipl.*, 226
— *tardigradus, Mic.*, 106
FLEMMING'S solution, 293, 300
Flies, diffusion of ferment by, 191
Fluoresceine staining method, 89
Fluorescens liquefaciens, Bac., 130
— *non-liquefaciens, Bac.*, 130
Fœtidus lactis, Bac., 185
— — *ozænæ, Bac.*, 266

Foods, methods of examining, 178, 189 ; moulds on, 193
Fowl cholera, Bac. of, 252
Fractional sterilisation, 18
FRÄNKEL, A., pneumococcus, 271
B., stain, 214
FRANKLAND, water-bacteria, 122
Freezing method, 79 ; in gum, 301
FRIEDLÄNDER, pneumobacillus, 273 ; stain, 213
Fruchthyphen, 103
Furuncles, staining of, 90
Fuscus limbatus, Bac., 174
— *Mic.*, 128

GABBET, method of staining tubercle bacilli, 75, 214
Gasoformans, Bac., 132
Gastric bacteria, 245
Gelatine, glycerine, 41 ; litmus, 41 ; meat extract peptone, 40 ; Miquel's, 42 ; peptone bouillon, 38 ; potato (Holtz), 42 ; serum, 49 ; urine, 42
Gemmation, multiplication by, 12
Generative organs, the, 222
Germs, sickle-shaped, 12
GIBBES, stain for tubercle bacilli, 215
Giganteus urethræ, Strept., 261
Gingivæ, Bac., 239
'Glacier Bacillus,' 130
Glanders, Bac. of, 218
'Glanders-nodules,' 218
Glycerine jelly, imbedding in, 81
Gonorrhœa, Mic. of, 201
Gracilis, Bac., 138
'La graisse,' 192
GRAM, decolorising method of, 76, 87 ; Günther's modification of, 77 ; Kühne's, 87 ; Weigert's, 78 ; dyes suited for, 77
Gregarinæ, 12
Ground-water, 155
Gum arabic, freezing in, 301 ; imbedding in, 81
GÜNTHER, methods of staining, 74, 87, 212, 277

Hæmatodes, Mic., 227
Hæmatoxyline, extract of, 30 ; stain, 293, 301
Hæmoglobinuria of oxen, 282
HAFFKINE, anti-cholera vaccination, 283
Hanging drop, examination in, 65
Hardening, methods of, 79 ; in alcohol, 80 ; chromic acid, 80 ; corrosive

x

sublimate, 81, 300; Flemming's solution, 300; picric acid, 80
'Hay bacillus.' See Bac. subtilis
Helicobacterium aerogenes, 251
'Hémoglobinurie bactérienne du bœuf,' 282
Hens' eggs, 50; white of, 37, 49
Herpes tonsurans, 232
HESSE, method of air-analysis, 100
High cultures, 62
Hollowed slides, 29
Hot-air steriliser, 15
Hot-water filtering funnel, 27
Hyphæ, 10
Hyphomycetæ, 10

Ianthinus, Bac., 138
Imbedding, 81; in celloidine, 82; in glycerine jelly, 81; in gum arabic, 81; in paraffine, 83
Immunity, 6
Impetigo contagiosa, 233
Impression preparations, 78
Incubators, 20; petroleum, Baumeyer's, 26
Indicus, Bac., 248
Indigoferus, Bac., 137
Indigogenus, Bac., 191
Infection, artificial, 92; by air-passages, 93; by digestive canal, 93; into eye, 95; intraperitoneal, 94, 284; intravenous, 94; subcutaneous, 94
Influenza, Bac. of, 2, 278
Inoculation. See Infection
Inspissator, serum, Koch's, 47, 48
Intestine, micro-organisms of, 249, 253
Iodine method, 89
Iodococcus magnus, I. parvus, I. vaginatus, 238
Islets, on plate cultures, 56

KAATZER, staining of tubercle bacilli, 211
Kephir, 188
Knob-moulds, 10
KOCH, comma-bacillus, 139; plate process, 54; serum-inspissator, 47; stain, 86; steam steriliser, 16; syringe, 31; tubercle bacillus, 206; tuberculine, 7, 8, 218
KOCH-EHRLICH method of staining tubercle bacilli, 73
KÜHNE, method of staining tubercle bacilli in sputum, 213; methyl blue, 86

Lacteus faviformis, Mic., 225
Lacticus, Bac., 179
Lactis aerogenes, Bac., 248
— erythrogenes, Bacter., 186
— fœtidus, Bac., 185
— pituitosi, Bac., 185
— viscosus, Bac., 184
'Laveran's sickles,' 171
Lepra, Bac. of, 229
— cells, 229
Leptothrix buccalis, 238, 255
Levelling stand, for plate cultures, 28
Light, action of, upon micro-organisms, 295; coloured, 297; white, 295; electric, 297; mode of action of, 298; natural precautions against, 299
Liodermos, Bac., 117
Liquefaciens magnus, Bac., 159
Litmus agar, 45; gelatine, 41
LÖFFLER, diphtheria bacillus, 240; methyl blue, 86, 241, 277
LÖSCH, amœba of, 288
Lutea, sarc., 109
Luteus, Dipl., 130
— Mic., 128

MACHINES, centrifugal, 31; Csokor's, 34; Gärtner's, 33; Stenbeck's, 31
Malariæ, Plasmodium, 169
Malignant pustule, 169
Malleus, 218. See Glanders
Mastitis, bovine, Mic. of, 187
Measurement, standards of, 65
Media, fixing, 84
— nutrient, 35; liquid, 35; solid, 37; turbidity in, 39
Megaterium, Bac., 189
Melanæmia, 170
Melanine, 170
Melochloros, Bac., 118
Membranaceus amethystinus, Bac.,137
Meningitidis intracellularis, Mic., 221
— purulentæ, Mic., 221
Meningitis, pneumococcus in, 271
Merismopedia, 4
Mesentericus fuscus, ruber, vulgatus, Bac., 116
Metabolism in bacteria, products of, 5; methods of extracting products (Brieger, Scholl, Roemer, Buchner, Koch), 7, 8, 218
Methyl blue, borax, Unna's, 90; carbolic, 88; and eosine (Chenzynsky's), 86, 278; Kühne's, 86; Löffler's, 86, 241, 277; Plehn's, 87
Metschnikoffi, Vibrio, 147

Micra. *See* Micro-millimeters
Micro-burners, 21
Micro-membrane filter, 123
Micro-millimeters, 65
Micro-organisms, classification of, 1, 2; examination of, 12; examination in fresh state and hanging drop, 65; in sections of tissue, 78; M. in disease, 92; isolation of, 57; by Cohen's method, 127; staining of, 67 *et seq.*; transmission of, to animals, 91
Microsporon furfur, 235
Milk, methods of examining, 179; micro-organisms in, 188
Milk-rice, 52-3
Miller, Spirillum, 242
MIQUEL, method of air-analysis, 98
Moist chambers, 28, 56
Molecular movement, 3
Molluscum contagiosum, protozoa in, 295
Monilia candida, 243
Monococci, 2
Mordants, 69
'Mother-of-vinegar,' 191
Mould, brown, 104, 156
Moulds, 9, 193; pathogenesis of, 194-5
Mouse septicæmia, 280
Mouth, the, micro-organisms of, 237, 244
Mucor corymbifer, 194; *M. mucedo*, 9, 194; *M. ramosus, M. rhizopodiformis*, 194
Mucorineæ, 10, 193
Multipediculosus, Bac., 118
Multiplication, modes of, 4, 194
Murisepticus, Bac., 280
Mycelium, 10, 103, 233
Mycoderma albicans, 243
Mycoides, Bac., 156
Mycosis fungoides, coccidia in, 295

NAIL, finger, 223
'Nail-culture,' 186, 248, 273
Nasal bacteria, 264, 268; secretion, 264
Nasalis, Mic., 265
— *Vibrio*, 268
Neapolitanus, Bac., 120, 147
NICOLLE, method of staining, 302
Nivalis, Bac., 130
NONIEWICZ, method of staining glanders bacilli, 220

Ochraceus, Bac., 138
Ochroleucus, Mic., 261
Œdematis maligni, Bac., 160
Oïdiaceæ, 11

Oïdium albicans, 243
— *lactis*, 10, 181
Onychomycosis, 223, 233
Organs and cavities of body, 222
Osteomyelitidis, Mic., 221
Otomycosis, 195
Ozæna, 266

PACKET-COCCI, 4
PAGET'S disease, protozoa in, 295
Pap, bread, 53; potato, 52
Paraffine, imbedding in, 83
Parasites, 1, 92, 286
Pediococcus cerevisiæ, 191
Pellicles, 4
Pemphigus acutus, Dipl., 227
Pencil mould, 11
'Pende's ulcer,' 221
Penicilliaceæ, 11, 103, 193
Penicillium glaucum, 10, 103
'Perlsucht,' 178
Permanent cultures, 63
PETRI, method of air-analysis, 102; capsules, 28, 58
Phlogosine, 113
Phosphorescence, 6, 133
Phosphorescens, Bac., 132; *indigenus*, 132; *indicus*, 133
Picrocarmine, 302
Pigment-forming bacteria, 5; effect of light on, 295, 299
Plasmodia, 12, 169, 282
Plate process, Koch's, 54; apparatus for, 28, 55; modifications of, 58, 59, 61
Platinum wires, 30
Plugs, cotton-wool, 16
Pneumobacillus Friedlænderi, 8, 273
Pneumoniæ, Diploc., 271
Pneumosepticus, Bac., 275
Polymorphic bacteria, 197
Potato bacillus, 116
Potatoes, preparation of, for culture-media, 51
Poultry typhoid, 252
Prodigiosus, Bac., 114
Proteines, 7
Proteus, 174; *P. capsulatus, P. hominis, P. mirabilis, P. Zenkeri*, 175; *P. vulgaris*, 175, 181
Protozoa, 12; in blood, 282; in carcinoma, 290; in molluscum contagiosum, 295; in mycosis fungoides, 295; in Paget's disease, 295; in sarcoma, 294; parasitic, 286; pathogenesis of, 286; staining of, 293. *See* also under Amœba, Coccidium, Plasmodium, &c.

x 2

Pseudodiphtheritic Bac., 242
Pseudonavicella, 12
Pseudopodia, 12
Psorosperms, 12
Ptomaines, 6, 173
Pulmonum, Sarc., 270
Pus, 196; microbes of, 196, 220; staining of, 221 *et passim*
Putrefactive processes, differences in, 173
Putrefying substances, analysis of, 173
Putrescine, 148
Putrificus coli, Bac., 251
Pyocyaneus, Bac., 8; α and β, 198-9, 223; effect of light on, 296
Pyocyanine, 199
Pyogenes, Staph., 109; *aureus*, 110; *albus, citreus*, 112
— *Strept.*, 113, 201

Radiatus, Bac., 158; *Mic.*, 105
'Rag-picker's disease,' 169
Ramosus, Bac., 133
'Rauschbrand,' 228
'Ray fungus,' 196. *See* Actinomyces
Reagents, bacteriological, 29
Respiratory passages, micro-organisms of, 269; R. tract, 264; infection through, 93
Rhinoscleroma, Bac. of, 266
Rice, milk, 52
Ringworm, 233
Roll cultures, Esmarch's, 58
Root bacillus, 133
Rosea, Sarc., 109, 187
Roseus, Mic., 107
Rubrum, Spir., 176
Rugula, Vibrio, 242
RUSSELL, method of staining, 293

Saccharomyces cerevisiæ, 11, 105; *S. ruber*, 188
Saffranine, staining with, 301
Saliva, 237, 238
Salivarius septicus, Bac., 238; *Mic.*, 238
Saprogenes, Bac., I., II., III., 175
Saprophytes, 1, 92
Sarcina, 2; *S. alba, S. aurantiaca*, 108; *S. aurea*, 270; *S. candida*, 108; *S. cerevisiæ*, 192; *S. lutea*, 109; *S. pulmonum*, 270; *S. rosea*, 109, 187; *S. ventriculi*, 246
Sarcoma, protozoa in, 294
Schizomycetes, 2, 4

Scissus, Bac., 159
Section-stainer, 82
Sections, cutting of, 79 *et seq.*, 301; examination of micro-organisms in, 68; staining of, 84
Sedimentation, methods of, 127, 216
Seed-organ, in moulds, 10
Segmentation, 170
Septicæmia, mouse, 280; puerperal, 238
Septicus, Bac., 160
— *vesicæ, Bac.*, 262
Serial preparations, 83, 84
Serum, blood, 47; modifications of, 48; sterilisation of, 19, 47
Skin, abscesses in, staining of, 90; micro-organisms of, 222; modes of staining, 223-4
Small-pox, protozoa in, 282
Smear preparations, 67
Smegma Bac., 206
Soil, examination of, 156; filtering action of, 156; micro-organisms of, 155, 172
'Soorpilz,' 243. *See* Thrush-fungus
SOYKA's plates, 29
Spectrum, experiments with, 297
Sphærococcus acidi lactici, 181
'Spider cells,' 148-9
Spinosus, Bac., 159
Spirilla, 2; atmospheric, 121; water, 153
Spiritus saponatus kalinus, 91
Spirochætæ, 2
Spirochæte dentium, 242
— *Obermeieri*, 281
Sporangium, 10
Spores, 4, 9, 103, 189, 291; effect of light on, 297, 299; staining of, 71; swarm, 12; terminal, 5, 163
Sporocysts, 12
Sporozoa, 12, 287
Sporulation, 4, 9, 291; arthrogenous, 5; endogenous, 5
Spring-water, 124
Sputum, 269; staining of, for tubercle bacilli, 210; pure cultures of tubercle bacilli from, 207
Staining, 67, 301; combination, 86; cover-glass preparations, 67; sections, 84. *See* also under Tubercle bacillus, Gonococcus, &c. &c.
— processes, Arens, 76, 214; Ehrlich, 74; Ehrlich-Biondi, 293; Fränkel, Gabbet, 75; Gibbes, 215; Gram, 76; Günther, 74, 212, 277; Koch and Ehrlich, 73, 210; Löffler, 86, 241,

277; Nicolle, 302; Pittion, Pfuhl and Petri, 75; Weichselbaum, 75; Weigert, 78, 89; Ziehl-Neelsen, 74, 213
Stains, 30; preparation of, 69
Staphylococci, 2, 4, 109
Staphylococcus cereus, 199; *S. pyogenes*, 109
Steam generator (Budenberg's), 17
- steriliser, Koch's, 16
Sterigmata, 11, 103
Sterilisation, 15; by chemical agents, 18; by combined methods, 19; by heat, 15; by steam, 16, 18; discontinuous or fractional, 18; of hands, 20; of instruments, 19
Steriliser, hot-air, 15; Koch's steam, 16
Stomach, micro-organisms of the, 245
STRAUSS and WÜRZ, method of analysing air, 101
Streptococci, 2, 4, 113
Streptococcus erysipelatis, 113; *S. giganteus urethræ*, 261; *S. pyogenes*, 113, 201; *S. septicus*, 165
Striatus albus, Bac., S. flavus, Bac., 266
Structural images, 68
Styrone method, 224
Subflavus, Dipl., 224
Subtiliformis, Bac., 256
Subtilis, Bac., 113
Subungual space, micro-organisms in, 223
Sulphureum, Bact., 262
Sulphydrogenus, Bac., 138
Sweat, 223
Swine erysipelas, Bac. of, 279
Sycosis, 231, 233
Sycosiferus fœtidus, Bac., 231
Symbiosis, 164
Symptomatic anthrax, Bac. of, 228
Syncyanus, Bac., 185
Syphilis, Bac., 205
Syringes, hypodermic, 31

TABLET-COCCI, 4
Temperatures, 21
Test-tubes, charging of, 40
Tetanus Bac., 163
Tetracocci, 2, 4
Tetragenus, Mic., 273
– – *concentricus, Mic.*, 256
— *mobilis ventriculi, Mic.*, 248
— *subflavus, Mic.*, 264
Texas fever, 282

Thallus, 10
Thermo-regulators, 21; Altmann's, 25; Bunsen's, 21; Gärtner's, 25; Meyer's, 24; Schenk's, 22
Thrush-fungus, 243
Tinea circinata, T. sycosis, 233; T. tonsurans, 232; T. versicolor, 234, 235
Tissues, the, influence of bacteria on, 6
Toxalbumins, 6
Toxines, 6
Toxopeptone, 7
Trachoma, Mic. of, 227
Transmission, 91; by air-passages, 93; digestive canal, 93; skin, 94; peritoneum, 94
Trichophyton macrosporon (megalosporon), T. microsporon, 236; *T. tonsurans*, 223, 232, 236
'Trouser-leg culture,' 145
Tuberculine, Koch's, 7, 8, 218
Tuberculocidine, 218
Tuberculosis, Bac., 206, 275; action of light on, 296; in birds, 209; preparing pure cultures of, 207–9; staining of, 73, 210; in sections, 217
Tubular buds, in moulds, 10
Tussis convulsivæ, Bac., 275
Tympanum, micro-organisms of, 244
TYNDALL, method of air-analysis, 102; of sterilisation, 18
Typhoid fever, Bac. of, 148; action of light on, 296
Typhus abdominalis, 148. See Typhoid
Tyrogenum, Spirillum, 183

Ulna, Bac., 239
Undula, Spirillum, 153
UNNA, examination of skin for micro-organisms, 90, 224
Ureæ, Bac., 260·; *Mic.*, 107
— *candidus, Staph., U. liquefaciens, Staph.*, 260
Urine, micro-organisms of, 258; yeasts and moulds in, 259
Urobacillus Freudenreichii, U. Maddoxii, 260
Urobacteria, 259

VACCINATION against cholera, Haffkine's, 283; principle of, 283; results of, 284
Vaccines, anti-cholera, preparation of, 283; attenuated, 285; exalted, 284

Vacuoles in cells, 12, 288
Vagina, micro-organisms of, 222-3
Vaginalis, Bac., 223, 228
Ventriculi, Sarcina, 246
Versicolor, Mic., 105
'Vibrion septique,' 160
Vibrios, 2, 4
— *aureus*, 121; *V. choleræ*, 139; *V. flavescens, V. flavus*, 121; *V. Metschnikoffi*, 147; *V. proteus*, 145
Vienna, water-supply of, 136
Vinegar ferment, 191
Violaceus, Bac., 132
Virulence, 6
Virus, exalted, 283
Viscosus, Mic., 192
— *butyri, Bac.*, 183; *V. cerevisiæ, Bac.*, 192; *V. lactis, Bac.*, 184; *V. sacchari, Bac.*, 193
Viticulosus, Mic., 106

WAFERS, cultivation on, 53
WARD, researches on the action of light on bacteria, 297-300
Warm chamber, 20. *See* Incubator
Water, analysis of, 122; W. bacteria, 122; W., drinking, 124; ground, 155; methods of examining, 125; Kirchner's, Pfuhl's, 126; other methods, 127; micro-organisms in, 122, 124, 153; spirilla in, 153; spring, 124; variations in, depending on source, 124
WEICHSELBAUM, method of staining tubercle bacilli, 75; pneumococcus, 271
WEIGERT, modification of Gram's method, 78, 89
WELZ, method of air-analysis, 99
Whooping cough, 275
WOLFFHÜGEL, counting apparatus, 125
'Wurzelbacillus,' 133

Xerosis, Bac., 232

YEAST, 12, 104; black y., 105; y.-cells, 12; pink y., white y., 105

ZIEHL'S solution, 70
— and NEELSEN'S method, 74
Zoogloea, 4
Zopfi, Bact., 251
Zuernianum, Bact., 137
Zygospores, 194

A LIST OF WORKS ON

MEDICINE, SURGERY
AND
GENERAL SCIENCE,

PUBLISHED BY

LONGMANS, GREEN & CO.,
LONDON, NEW YORK, AND BOMBAY.

Medical and Surgical Works.

ASHBY. NOTES ON PHYSIOLOGY FOR THE USE OF STUDENTS PREPARING FOR EXAMINATION. By HENRY ASHBY, M.D. Lond., F.R.C.P., Physician to the General Hospital for Sick Children, Manchester; formerly Demonstrator of Physiology, Liverpool School of Medicine. Sixth Edition, thoroughly revised. With 141 Illustrations. Fcap. 8vo, price 5s.

ASHBY AND WRIGHT. THE DISEASES OF CHILDREN, MEDICAL AND SURGICAL. By HENRY ASHBY, M.D. Lond., F.R.C.P., Physician to the General Hospital for Sick Children, Manchester; Lecturer and Examiner in Diseases of Children in the Victoria University; and G. A. WRIGHT, B.A., M.B. Oxon., F.R.C.S. Eng., Assistant Surgeon to the Manchester Royal Infirmary and Surgeon to the Children's Hospital; formerly Examiner in Surgery in the University of Oxford. Enlarged and Improved Edition. With 192 Illustrations. 8vo, price 25s.

BENNETT.—*WORKS by WILLIAM H. BENNETT, F.R.C.S.,*
Surgeon to St. George's Hospital; Member of the Board of Examiners, Royal College of Surgeons of England.

CLINICAL LECTURES ON VARICOSE VEINS OF THE LOWER EXTREMITIES. With 3 Plates. 8vo. 6s.

ON VARICOCELE: A PRACTICAL TREATISE. With 4 Tables and a Diagram. 8vo. 5s.

CLINICAL LECTURES ON ABDOMINAL HERNIA: chiefly in relation to Treatment, including the Radical Cure. With 12 Diagrams in the Text. 8vo. 8s. 6d.

WORKS ON MEDICINE, SURGERY &c.

CLARKE. POST-MORTEM EXAMINATIONS IN MEDICO-LEGAL AND ORDINARY CASES. With Special Chapters on the Legal Aspects of Post-Mortems, and on Certificates of Death. By J. JACKSON CLARKE, M.B. (Lond.), F.R.C.S., Assistant-Surgeon to the North-West London Hospital, Pathologist and Curator of the Museum at St. Mary's Hospital. Fcp. 8vo, 2s. 6d.

COATS. A MANUAL OF PATHOLOGY. By JOSEPH COATS, M.D., Professor of Pathology in the University of Glasgow. Third Edition. Revised throughout. With 507 Illustrations. 8vo, 31s. 6d.

COOKE.—WORKS by THOMAS COOKE, F.R.C.S. Eng., B.A., B.Sc., M.D. Paris, Senior Assistant Surgeon to the Westminster Hospital.

TABLETS OF ANATOMY. Being a Synopsis of Demonstrations given in the Westminster Hospital Medical School in the years 1871–75. Tenth Thousand, being a selection of the Tablets believed to be most useful to Students generally. Post 4to, price 10s. 6d.

APHORISMS IN APPLIED ANATOMY AND OPERATIVE SURGERY. Including 100 Typical *vivâ voce* Questions on Surface Marking, &c. Crown 8vo, 3s. 6d.

DISSECTION GUIDES. Aiming at Extending and Facilitating such Practical Work in Anatomy as will be specially useful in connection with an ordinary Hospital Curriculum. 8vo, 10s. 6d.

DICKINSON.—WORKS by W. HOWSHIP DICKINSON, M.D. Cantab., F.R.C.P., Physician to, and Lecturer on Medicine at, St. George's Hospital; Consulting Physician to the Hospital for Sick Children.

ON RENAL AND URINARY AFFECTIONS. Complete in Three Parts, 8vo, with 12 Plates and 122 Woodcuts. Price £3 4s. 6d. cloth.

** The Parts can also be had separately, each complete in itself, as follows :—

PART I.—*Diabetes*, price 10s. 6d. sewed, 12s. cloth.

,, II.—*Albuminuria*, price £1 sewed, £1 1s. cloth.

,, III.—*Miscellaneous Affections of the Kidneys and Urine*, price £1 10s. sewed, £1 11s. 6d. cloth.

THE TONGUE AS AN INDICATION OF DISEASE; being the Lumleian Lectures delivered at the Royal College of Physicians in March, 1888. 8vo, price 7s. 6d.

THE HARVEIAN ORATION ON HARVEY IN ANCIENT AND MODERN MEDICINE. Crown 8vo, 2s. 6d.

OCCASIONAL PAPERS ON MEDICAL SUBJECTS, 1855–1896. 8vo, 12s.

DUCKWORTH. THE SEQUELS OF DISEASE: being the Lumleian Lectures delivered in the Royal College of Physicians, 1896. Together with Observations on Prognosis in Disease. By Sir DYCE DUCKWORTH, M.D., LL.D., Fellow and Treasurer of the Royal College of Physicians, &c. 8vo, 10*s*. 6*d*.

ERICHSEN.—THE SCIENCE AND ART OF SURGERY; A TREATISE ON SURGICAL INJURIES, DISEASES, AND OPERATIONS. By Sir JOHN ERIC ERICHSEN, Bart., F.R.S., LL.D. (Edin.), Hon. M. Ch. and F.R.C.S. (Ireland), Surgeon Extraordinary to H.M. the Queen; President of University College, London; Fellow and Ex-President of the Royal College of Surgeons of England; Emeritus Professor of Surgery in University College; Consulting-Surgeon to University College Hospital, and to many other Medical Charities. Tenth Edition. Revised by the late MARCUS BECK, M.S. & M.B. (Lond.), F.R.C.S., Surgeon to University College Hospital, and Professor of Surgery in University College, London; and by RAYMOND JOHNSON, M.B. & B.S. (Lond.), F.R.C.S., Assistant Surgeon to University College Hospital, &c. Illustrated by nearly 1,000 Engravings on Wood. 2 Vols. royal 8vo, 48*s*.

FRANKLAND. MICRO-ORGANISMS IN WATER, THEIR SIGNIFICANCE, IDENTIFICATION, AND REMOVAL. Together with an Account of the Bacteriological Methods Involved in their Investigation. Specially Designed for the Use of those connected with the Sanitary Aspects of Water Supply. By Professor PERCY FRANKLAND, Ph.D., B.Sc. (Lond.), F.R.S., Fellow of the Chemical Society; and Mrs. PERCY FRANKLAND, Joint Author of "Studies on Some New Micro-Organisms Obtained from Air." With 2 Plates and numerous Diagrams. 8vo. 16*s. net*.

GARROD.—*WORKS by Sir ALFRED BARING GARROD, M.D., F.R.S., &c.; Physician Extraordinary to H.M. the Queen; Consulting Physician to King's College Hospital; late Vice-President of the Royal College of Physicians.*

A TREATISE ON GOUT AND RHEUMATIC GOUT (RHEUMATOID ARTHRITIS). Third Edition, thoroughly revised and enlarged; with 6 Plates, comprising 21 Figures (14 Coloured), and 27 Illustrations engraved on Wood. 8vo, price 21*s*.

THE ESSENTIALS OF MATERIA MEDICA AND THERAPEUTICS. The Thirteenth Edition, revised and edited, under the supervision of the Author, by NESTOR TIRARD, M.D. Lond., F.R.C.P., Professor of Materia Medica and Therapeutics in King's College, London, &c. Crown 8vo, price 12*s*. 6*d*.

GRAY. ANATOMY, DESCRIPTIVE AND SURGICAL. By HENRY GRAY, F.R.S., late Lecturer on Anatomy at St. George's Hospital. The Thirteenth Edition, re-edited by T. PICKERING PICK, Surgeon to St. George's Hospital; Inspector of Anatomy in England and Wales; late Member of the Court of Examiners, Royal College of Surgeons of England. With 636 large Woodcut Illustrations, a large proportion of which are Coloured, the Arteries being coloured red, the Veins blue, and the Nerves yellow. The attachments of the muscles to the bones, in the section on Osteology, are also shown in coloured outline. Royal 8vo, price 36s.

HALLIBURTON.—*WORKS by W. D. HALLIBURTON, M.D.,*
F.R.S., M.R.C.P., Professor of Physiology in King's College, London; Lecturer on Physiology at the London School of Medicine for Women.

A TEXT-BOOK OF CHEMICAL PHYSIOLOGY AND PATHOLOGY. With 104 Illustrations. 8vo, 28s.

ESSENTIALS OF CHEMICAL PHYSIOLOGY. Second Edition. 8vo, 5s.

*** This is a book suitable for medical students. It treats of the subject in the same way as Prof. SCHÄFER's " Essentials " treats of Histology. It contains a number of elementary and advanced practical lessons, followed in each case by a brief descriptive account of the facts related to the exercises which are intended to be performed by each member of the class.

LANG.—THE METHODICAL EXAMINATION OF THE EYE. Being Part I. of a Guide to the Practice of Ophthalmology for Students and Practitioners. By WILLIAM LANG, F.R.C.S. Eng., Surgeon to the Royal London Ophthalmic Hospital, Moorfields, &c. With 15 Illustrations. Crown 8vo, 3s. 6d.

LIVEING.—*WORKS by ROBERT LIVEING, M.A. & M.D. Cantab.,*
F.R.C.P. Lond., &c., Physician to the Department for Diseases of the Skin at the Middlesex Hospital, &c.

HANDBOOK ON DISEASES OF THE SKIN. With especial reference to Diagnosis and Treatment. Fifth Edition, revised and enlarged. Fcap. 8vo, price 5s.

ELEPHANTIASIS GRÆCORUM, OR TRUE LEPROSY; Being the Goulstonian Lectures for 1873. Cr. 8vo, 4s. 6d.

LONGMORE.—*WORKS by Surgeon-General Sir T. LONGMORE* (*Retired*), *C.B., F.R.C.S., late Professor of Military Surgery in the Army Medical School; Officer of the Legion of Honour.*

THE ILLUSTRATED OPTICAL MANUAL; OR, HANDBOOK OF INSTRUCTIONS FOR THE GUIDANCE OF SURGEONS IN TESTING QUALITY AND RANGE OF VISION, AND IN DISTINGUISHING AND DEALING WITH OPTICAL DEFECTS IN GENERAL. Illustrated by 74 Drawings and Diagrams by Inspector-General Dr. MACDONALD, R.N., F.R.S., C.B. Fourth Edition. 8vo, price 14s.

GUNSHOT INJURIES. Their History, Characteristic Features, Complications, and General Treatment; with Statistics concerning them as they have been met with in Warfare. With 78 Illustrations. 8vo, price 31s. 6d.

LUFF. TEXT-BOOK OF FORENSIC MEDICINE AND TOXICOLOGY. By ARTHUR P. LUFF, M.D., B.Sc. (Lond.), Physician in Charge of Out-Patients and Lecturer on Medical Jurisprudence and Toxicology in St. Mary's Hospital; Examiner in Forensic Medicine in the University of London; External Examiner in Forensic Medicine in the Victoria University; Official Analyst to the Home Office. With numerous Illustrations. 2 vols., crown 8vo, 24s.

MUNK. THE LIFE OF SIR HENRY HALFORD, Bart., G.C.H., M.D., F.R.S., President of the Royal College of Physicians, Physician to George III., George IV., William IV., and to Her Majesty Queen Victoria. By William MUNK, M.D., F.S.A., Fellow and late Vice-President of the Royal College of Physicians of London. With 2 Portraits. 8vo, 12s. 6d.

NEWMAN. ON THE DISEASES OF THE KIDNEY AMENABLE TO SURGICAL TREATMENT. Lectures to Practitioners. By DAVID NEWMAN, M.D., Surgeon to the Western Infirmary Out-Door Department; Pathologist and Lecturer on Pathology at the Glasgow Royal Infirmary; Examiner in Pathology in the University of Glasgow; Vice-President Glasgow Pathological and Clinical Society. 8vo, price 16s.

WORKS ON MEDICINE, SURGERY &c.

NOTTER AND FIRTH. HYGIENE. By J. L. NOTTER, M.A., M.D., Fellow and Member of Council of the Sanitary Institute of Great Britain, Examiner in Hygiene, Science and Art Department, etc.; and R. H. FIRTH, F.R.C.S., Assistant Examiner in Hygiene, Science and Art Department. With 93 Illustrations. Crown 8vo, 3s. 6d.

OWEN. A MANUAL OF ANATOMY FOR SENIOR STUDENTS. By EDMUND OWEN, M.B., F.R.S.C., Senior Surgeon to the Hospital for Sick Children, Great Ormond Street, Surgeon to St. Mary's Hospital, London, and co-Lecturer on Surgery, late Lecturer on Anatomy in its Medical School. With 210 Illustrations. Crown 8vo, price 12s. 6d.

POOLE. COOKERY FOR THE DIABETIC. By W. H. and Mrs. POOLE. With Preface by Dr. PAVY. Fcap. 8vo. 2s. 6d.

POORE. ESSAYS ON RURAL HYGIENE. By GEORGE VIVIAN POORE, M.D., F.R.C.P. Crown 8vo, 6s. 6d.

QUAIN. A DICTIONARY OF MEDICINE; Including General Pathology, General Therapeutics, Hygiene, and the Diseases of Women and Children. By Various Writers. Edited by RICHARD QUAIN, Bart., M.D.Lond., LL.D.Edin. (Hon.) F.R.S., Physician Extraordinary to H.M. the Queen, President of the General Medical Council, Member of the Senate of the University of London, &c. Assisted by FREDERICK THOMAS ROBERTS, M.D.Lond., B.Sc., Fellow of the Royal College of Physicians, Fellow of University College, Professor of Materia Medica and Therapeutics, University College, &c.; and J. MITCHELL BRUCE, M.A.Abdn., M.D.Lond., Fellow of the Royal College of Physicians of London, Physician and Lecturer on the Principles and Practice of Medicine, Charing Cross Hospital, &c. New Edition, Revised throughout and Enlarged. In 2 Vols. medium 8vo, cloth, red edges, price 40s. net.

QUAIN. QUAIN'S (JONES) ELEMENTS OF ANATOMY. The Tenth Edition. Edited by EDWARD ALBERT SCHÄFER, F.R.S., Professor of Physiology and Histology in University College, London; and GEORGE DANCER THANE, Professor of Anatomy in University College, London. In 3 Vols.

⁎ The several parts of this work form COMPLETE TEXT-BOOKS OF THEIR RESPECTIVE SUBJECTS. They can be obtained separately as follows:—

VOL. I., PART I. EMBRYOLOGY. By E. A. SCHÄFER, F.R.S. With 200 Illustrations. Royal 8vo, 9s.

VOL. I., PART II. GENERAL ANATOMY OR HISTOLOGY. By E. A. SCHÄFER, F.R.S. With 291 Illustrations. Royal 8vo, 12s. 6d.

VOL. II., PART I. OSTEOLOGY. By G. D. THANE. With 168 Illustrations. Royal 8vo, 9s.

VOL. II., PART II. ARTHROLOGY — MYOLOGY — ANGEIOLOGY. By G. D. THANE. With 255 Illustrations. Royal 8vo, 18s.

VOL. III., PART I. THE SPINAL CORD AND BRAIN. By E. A. SCHÄFER, F.R.S. With 139 Illustrations. Royal 8vo, 12s. 6d.

VOL. III., PART II. THE NERVES. By G. D. THANE. With 102 Illustrations. Royal 8vo, 9s.

VOL. III., PART III. THE ORGANS OF THE SENSES. By E. A. SCHÄFER, F.R.S. With 178 Illustrations. Royal 8vo, 9s.

VOL. III., PART IV. SPLANCHNOLOGY. By E. A. SCHÄFER, F.R.S., and JOHNSON SYMINGTON, M.D. With 337 Illustrations. Royal 8vo, 16s.

APPENDIX. SUPERFICIAL AND SURGICAL ANATOMY. By Professor G. D. THANE and Professor R. J. GODLEE, M.S. With 29 Illustrations. Royal 8vo, 6s. 6d.

SCHÄFER. THE ESSENTIALS OF HISTOLOGY: Descriptive and Practical. For the Use of Students. By E. A. SCHÄFER, F.R.S., Jodrell Professor of Physiology in University College, London; Editor of the Histological Portion of Quain's "Anatomy." Illustrated by more than 300 Figures, many of which are new. Fourth Edition, Revised and Enlarged. 8vo, 7s. 6d. (Interleaved, 10s.)

SCHENK. MANUAL OF BACTERIOLOGY. For Practitioners and Students. With especial reference to Practical Methods. By Dr. S. L. SCHENK, Professor (Extraordinary) in the University of Vienna. Translated from the German, with an Appendix, by W. R. DAWSON, B.A., M.D., Univ. Dub.; late University Travelling Prizeman in Medicine. With 100 Illustrations, some of which are coloured. 8vo, 10s. net.

SMALE AND COLYER. DISEASES AND INJURIES OF THE TEETH, including Pathology and Treatment : a Manual of Practical Dentistry for Students and Practitioners. By MORTON SMALE, M.R.C.S., L.S.A., L.D.S., Dental Surgeon to St. Mary's Hospital, Dean of the School, Dental Hospital of London, &c. ; and J. F. COLYER, L.R.C.P., M.R.C.S., L.D.S., Assistant Dental Surgeon to Charing Cross Hospital, and Assistant Dental Surgeon to the Dental Hospital of London. With 334 Illustrations. Large Crown 8vo, 15s.

SMITH (H. F.). THE HANDBOOK FOR MIDWIVES. By HENRY FLY SMITH, B.A., M.B. Oxon., M.R.C.S. Second Edition. With 41 Woodcuts. Crown 8vo, price 5s.

STEEL.—*WORKS by JOHN HENRY STEEL, F.R.C.V.S., F.Z.S., A.V.D., late Professor of Veterinary Science and Principal of Bombay Veterinary College.*

A TREATISE ON THE DISEASES OF THE DOG; being a Manual of Canine Pathology. Especially adapted for the use of Veterinary Practitioners and Students. 88 Illustrations. 8vo, 10s. 6d.

A TREATISE ON THE DISEASES OF THE OX; being a Manual of Bovine Pathology. Especially adapted for the use of Veterinary Practitioners and Students. 2 Plates and 117 Woodcuts. 8vo, 15s.

A TREATISE ON THE DISEASES OF THE SHEEP; being a Manual of Ovine Pathology for the use of Veterinary Practitioners and Students. With Coloured Plate, and 99 Woodcuts. 8vo, 12s.

OUTLINES OF EQUINE ANATOMY; a Manual for the use of Veterinary Students in the Dissecting Room. Crown 8vo, 7s. 6d.

"STONEHENGE." THE DOG IN HEALTH AND DISEASE. By "STONEHENGE." With 84 Wood Engravings. Square crown 8vo, 7s. 6d.

THORNTON. HUMAN PHYSIOLOGY. By JOHN THORNTON, M.A., Author of "Elementary Physiography," "Advanced Physiography," &c. With 267 Illustrations, some of which are Coloured. Crown 8vo, 6s.

TIRARD. DIPHTHERIA AND ANTITOXIN. By NESTOR TIRARD, M.D. Lond., Fellow of the Royal College of Physicians ; Fellow of King's College, London ; Professor of Materia Medica and Therapeutics at King's College ; Physician to King's College Hospital ; and Senior Physician to the Evelina Hospital for Sick Children. 8vo, 7s. 6d.

WALLER. AN INTRODUCTION TO HUMAN PHYSIOLOGY.
By AUGUSTUS D. WALLER, M.D., Lecturer on Physiology at St. Mary's Hospital Medical School, London ; late External Examiner at the Victorian University. Second Edition, Revised. With 305 Illustrations. 8vo, 18s.

WALLER AND SYMES. EXERCISES IN PRACTICAL PHYSIOLOGY. By AUGUSTUS D. WALLER, M.D., F.R.S. Part I. ELEMENTARY PHYSIOLOGICAL CHEMISTRY. By AUGUSTUS D. WALLER, M.D., and W. LEGGE SYMES. 8vo, sewed, 1s. net.

WEICHSELBAUM. THE ELEMENTS OF PATHOLOGICAL HISTOLOGY, With Special Reference to Practical Methods. By Dr. ANTON WEICHSELBAUM, Professor of Pathology in the University of Vienna. Translated by W. R. DAWSON, M.D. (Dub.), Demonstrator of Pathology in the Royal College of Surgeons, Ireland, late Medical Travelling Prizeman of Dublin University, &c. With 221 Figures, partly in Colours, a Chromo-lithographic Plate, and 7 Photographic Plates. Royal 8vo, 21s. net.

WILKS AND MOXON. LECTURES ON PATHOLOGICAL ANATOMY. By SAMUEL WILKS, M.D., F.R.S., Consulting Physician to, and formerly Lecturer on Medicine and Pathology at, Guy's Hospital, and the late WALTER MOXON, M.D., F.R.C.P., Physician to, and some time Lecturer on Pathology at, Guy's Hospital. Third Edition, thoroughly Revised. By SAMUEL WILKS, M.D., LL.D., F.R.S. 8vo, 18s.

YOUATT.—*WORKS by WILLIAM YOUATT.*

THE HORSE. Revised and Enlarged by W. WATSON, M.R.C.V.S. With 52 Woodcuts. 8vo, 7s. 6d.

THE DOG. Revised and Enlarged. With 33 Woodcuts. 8vo, 6s.

General Scientific Works.

BENNETT AND MURRAY. A HANDBOOK OF CRYPTOGAMIC BOTANY. By A. W. BENNETT, M.A., B.Sc., F.L.S., and GEORGE R. MILNE MURRAY, F.L.S. With 378 Illustrations. 8vo, 16s.

CLERKE. THE SYSTEM OF THE STARS. By AGNES M. CLERKE, Author of "A History of Astronomy during the Nineteenth Century." With 6 Plates and Numerous Illustrations. 8vo, 21s.

CLODD.—*WORKS by EDWARD CLODD, Author of "The Childhood of the World," &c.*

THE STORY OF CREATION. A Plain Account of Evolution. With 77 Illustrations. Crown 8vo, 3s. 6d.

A PRIMER OF EVOLUTION: being a Popular Abridged Edition of "The Story of Creation." With Illustrations. Fcp. 8vo, 1s. 6d.

CROOKES. SELECT METHODS IN CHEMICAL ANALYSIS (chiefly Inorganic). By W. CROOKES, F.R.S., V.P.C.S., Editor of "The Chemical News." Third Edition, re-written and enlarged. Illustrated with 67 Woodcuts. 8vo, 21s. net.

CULLEY. A HANDBOOK OF PRACTICAL TELEGRAPHY. By R. S. CULLEY, M.I.C.E., late Engineer-in-Chief of Telegraphs to the Post Office. With 135 Woodcuts and 17 Plates, 8vo, 16s.

DU BOIS. THE MAGNETIC CIRCUIT IN THEORY AND PRACTICE. By Dr. H. DU BOIS, Privat-docent in the University of Berlin. Translated from the German by Dr. E. ATKINSON. With 94 Illustrations. 8vo, 12s. net.

EBERT. MAGNETIC FIELDS OF FORCE: An Exposition of the Phenomena of Magnetism, Electromagnetism, and Induction, based on the Conception of Lines of Force. By H. EBERT, Professor of Physics in the University of Kiel. Translated by C. V. BURTON, D.Sc. Part I. With 93 Illustrations. 8vo, 10s. 6d. net.

GANOT. ELEMENTARY TREATISE ON PHYSICS; Experimental and Applied, for the use of Colleges and Schools. Translated and edited from GANOT's *Eléments de Physique* (with the Author's sanction) by E. ATKINSON, Ph.D., F.C.S., formerly Professor of Experimental Science, Staff College, Sandhurst. Fourteenth Edition, revised and enlarged, with 9 Coloured Plates and 1,028 Woodcuts. Large crown 8vo, 15s.

NATURAL PHILOSOPHY FOR GENERAL READERS AND YOUNG PERSONS; Translated and Edited from GANOT's *Cours Élémentaire de Physique* (with the Author's sanction) by E. ATKINSON, Ph.D., F.C.S. Eighth Edition, carefully revised; with 7 Plates, 624 Woodcuts, and an Appendix of Questions. Crown 8vo, 7s. 6d.

PUBLISHED BY LONGMANS, GREEN & CO. 11

GOODEVE.—*WORKS by T. M. GOODEVE, M.A., Barrister-at-Law; formerly Professor of Mechanics at the Normal School of Science and the Royal School of Mines.*

PRINCIPLES OF MECHANICS. New Edition, re-written and enlarged. With 253 Woodcuts and numerous Examples. Crown 8vo, 6s.

THE ELEMENTS OF MECHANISM. New Edition, re-written and enlarged. With 342 Woodcuts. Crown 8vo, 6s.

A MANUAL OF MECHANICS: an Elementary Text-Book for Students of Applied Mechanics. With 138 Illustrations and Diagrams, and 141 Examples taken from the Science Department Examination Papers, with Answers. Fcp. 8vo, 2s. 6d.

HELMHOLTZ.—*WORKS by HERMANN L. F. HELMHOLTZ, M.D., late Professor of Physics in the University of Berlin.*

ON THE SENSATIONS OF TONE AS A PHYSIOLOGICAL BASIS FOR THE THEORY OF MUSIC. Second English Edition; with numerous additional Notes, and a new Additional Appendix, bringing down information to 1885, and specially adapted to the use of Musical Students. By ALEXANDER J. ELLIS, B.A., F.R.S., F.S.A., &c., formerly Scholar of Trinity College, Cambridge. With 68 Figures engraved on Wood, and 42 Passages in Musical Notes. Royal 8vo, 28s.

POPULAR LECTURES ON SCIENTIFIC SUBJECTS. With 68 Woodcuts. 2 Vols. crown 8vo, 3s. 6d. each.

HERSCHEL. OUTLINES OF ASTRONOMY. By Sir JOHN F. W. HERSCHEL, Bart., K.H., &c., Member of the Institute of France. Twelfth Edition, with 9 Plates, and numerous Diagrams. 8vo, 12s.

HUDSON AND GOSSE. THE ROTIFERA OR 'WHEEL ANIMALCULES.' By C. T. HUDSON, LL.D., and P. H. GOSSE, F.R.S. With 30 Coloured and 4 Uncoloured Plates. In 6 Parts. 4to, price 10s. 6d. each; Supplement, 12s. 6d. Complete in Two Volumes, with Supplement, 4to, £4 4s.

*** The Plates in the Supplement contain figures of almost all the Foreign Species, as well as of the British Species, that have been discovered since the original publication of Vols. I. and II.

JOUBERT. ELEMENTARY TREATISE ON ELECTRICITY AND MAGNETISM. Founded on JOUBERT'S "*Traité Élémentaire d'Électricité.*" By G. C. FOSTER, F.R.S., Quain Professor of Physics in University College, London; and E. ATKINSON, Ph.D., formerly Professor of Experimental Science in the Staff College. With 381 Illustrations. Crown 8vo, 7s. 6d.

KOLBE. A SHORT TEXT-BOOK OF INORGANIC CHEMISTRY. By Dr. HERMANN KOLBE, late Professor of Chemistry in the University of Leipzig. Translated and Edited by T. S. HUMPIDGE, Ph.D., B.Sc. (Lond.), late Professor of Chemistry and Physics in the University College of Wales, Aberystwyth. New Edition. Revised by H. LLOYD-SNAPE, Ph.D., D.Sc. (Lond.). Professor of Chemistry in the University College of Wales, Aberystwyth. With a Coloured Table of Spectra and 66 Woodcuts. Crown 8vo, 8s. 6d.

LARDEN. ELECTRICITY FOR PUBLIC SCHOOLS AND COLLEGES. With numerous Questions and Examples with Answers, and 214 Illustrations and Diagrams. By W. LARDEN, M.A. Crown 8vo, 6s.

LEWIS. PAPERS AND NOTES ON THE GENESIS AND MATRIX OF THE DIAMOND. By the late HENRY CARVILL LEWIS, M.A., F.G.S., Professor of Mineralogy in the Academy of Natural Sciences, Philadelphia, Professor of Geology in Haverford College, U.S.A. Edited from his unpublished MSS. by Professor T. G. BONNEY, D.Sc., LL.D., F.R.S., &c. [*In the press.*

LINDLEY AND MOORE. THE TREASURY OF BOTANY, OR POPULAR DICTIONARY OF THE VEGETABLE KINGDOM: with which is incorporated a Glossary of Botanical Terms. Edited by J. LINDLEY, M.D., F.R.S., and T. MOORE, F.L.S. With 20 Steel Plates, and numerous Woodcuts. 2 Parts, fcp. 8vo, price 12s.

LOWELL. MARS. By PERCIVAL LOWELL, Fellow American Academy, Member Royal Asiatic Society, Great Britain and Ireland, &c. With 24 Plates. 8vo, 12s. 6d.

*** The book is written in a style suitable for the general reader, and the most recent speculations as to the planet being inhabited, the possible canals, oases, &c., are discussed.

MARTIN. NAVIGATION AND NAUTICAL ASTRONOMY. Compiled by Staff-Commander W. R. MARTIN, R.N., Instructor in Surveying, Navigation, and Compass Adjustment; Lecturer on Meteorology at the Royal Naval College, Greenwich. Sanctioned for use in the Royal Navy by the Lords Commissioners of the Admiralty. Royal 8vo, 18s.

MENDELÉEFF. THE PRINCIPLES OF CHEMISTRY. By D. MENDELÉEFF, Professor of Chemistry in the University of St. Petersburg. Translated by GEORGE KAMENSKY, A.R.S.M. of the Imperial Mint, St. Petersburg, and Edited by A. J. GREENAWAY, F.I.C., Sub-Editor of the Journal of the Chemical Society. With 97 Illustrations. 2 Vols. 8vo, 36s.

MEYER. OUTLINES OF THEORETICAL CHEMISTRY. By LOTHAR MEYER, Professor of Chemistry in the University of Tübingen. Translated by Professors P. PHILLIPS BEDSON, D.Sc., and W. CARLETON WILLIAMS, B.Sc. 8vo, 9s.

MITCHELL. MANUAL OF PRACTICAL ASSAYING. By JOHN MITCHELL, F.C.S. Sixth Edition. Edited by W. CROOKES, F.R.S. With 201 Woodcuts. 8vo, 31s. 6d.

MORGAN. ANIMAL BIOLOGY. An Elementary Text Book. By C. LLOYD MORGAN, Principal of University College, Bristol. With numerous Illustrations. Crown 8vo, 8s. 6d.

MOSSO. FEAR. By ANGELO MOSSO. Translated from the Fifth Edition of the Italian by E. LOUGH and F. KIESOW. With 8 Illustrations. Crown 8vo, 7s. 6d.

** This book deals with much more than is conveyed by the title. It is, in fact, a series of essays on the expression of the emotions, dealing more especially with the painful emotions. Although the subject is treated in a measure scientifically, i.e., physiologically, the book is not intended solely for the scientific public.

MUIR. THE ALCHEMICAL ESSENCE AND THE CHEMICAL ELEMENT: an Episode in the Quest of the Unchanging. By M. M. PATTISON MUIR, Fellow of Gonville and Caius College, Cambridge. 8vo. 4s. 6d.

OSTWALD. SOLUTIONS. By W. OSTWALD, Professor of Chemistry in the University of Leipzig. Being the Fourth Book, with some additions, of the Second Edition of Ostwald's "Lehrbuch der Allgemeinen Chemie." Translated by M. M. PATTISON MUIR, Professor of Gonville and Caius College, Cambridge. 8vo, 10s. 6d.

PAYEN. INDUSTRIAL CHEMISTRY; A Manual for use in Technical Colleges or Schools, also for Manufacturers and others, based on a Translation of Stohmann and Engler's German Edition of PAYEN'S *Précis de Chimie Industrielle*. Edited and supplemented with Chapters on the Chemistry of the Metals, &c., by B. H. PAUL, Ph.D. With 698 Woodcuts. 8vo, 42s.

PROCTOR.—*WORKS by RICHARD A. PROCTOR.*

OLD AND NEW ASTRONOMY. By RICHARD A. PROCTOR and A. COWPER RANYARD. With 31 Plates and 472 Illustrations. Text. 4to, 21s.

LIGHT SCIENCE FOR LEISURE HOURS; Familiar Essays on Scientific Subjects, Natural Phenomena, &c. 3 Vols. Crown 8vo, 5s. each.

THE ORBS AROUND US; a Series of Essays on the Moon and Planets, Meteors, and Comets. With Chart and Diagrams. Crown 8vo, 3s. 6d.

OTHER WORLDS THAN OURS; The Plurality of Worlds Studied under the Light of Recent Scientific Researches. With 14 Illustrations. Crown 8vo, 3s. 6d.

[*Continued.*

GENERAL SCIENTIFIC WORKS

PROCTOR.—*WORKS by RICHARD A. PROCTOR—continued.*

THE MOON; her Motions, Aspects, Scenery, and Physical Condition. With Plates, Charts, Woodcuts, and Lunar Photographs. Crown 8vo, 5s.

UNIVERSE OF STARS; Presenting Researches into and New Views respecting the Constitution of the Heavens. With 22 Charts and 22 Diagrams. 8vo. 10s. 6d.

LARGER STAR ATLAS for the Library, in 12 Circular Maps, with Introduction and 2 Index Pages. Folio, 15s., or Maps only, 12s. 6d.

NEW STAR ATLAS for the Library, the School, and the Observatory, in 12 Circular Maps (with 2 Index Plates). Crown 8vo, 5s.

OTHER SUNS THAN OURS: a Series of Essays on Suns—Old, Young, and Dead. With other Science Gleanings, Two Essays on Whist, and Correspondence with Sir John Herschel. With 9 Star-Maps and Diagrams. Crown 8vo, 3s. 6d.

HALF-HOURS WITH THE TELESCOPE: a Popular Guide to the Use of the Telescope as a Means of Amusement and Instruction. With 7 Plates. Fcap. 8vo, 2s. 6d.

THE SOUTHERN SKIES: a Plain and Easy Guide to the Constellations of the Southern Hemisphere. Showing in 12 Maps the position of the principal Star-Groups night after night throughout the year. With an Introduction and a separate Explanation of each Map. True for every Year. 4to, 5s.

HALF-HOURS WITH THE STARS: a Plain and Easy Guide to the Knowledge of the Constellations. Showing in 12 Maps the position of the principal Star-Groups night after night throughout the Year. With Introduction and a separate Explanation of each Map. True for every Year. 4to, 3s. 6d.

THE STARS IN THEIR SEASONS. An Easy Guide to a Knowledge of the Star Groups, in 12 Large Maps. Imperial 8vo, 5s.

OUR PLACE AMONG INFINITIES: a Series of Essays contrasting our Little Abode i Space and Time with the Infinities around Us. Crown 8vo, 3s. 6d.

ROUGH WAYS MADE SMOOTH. Familiar Essays on Scientific Subjects. Crown 8vo, 3s. 6d.

THE EXPANSE OF HEAVEN. Essays on the Wonders of the Firmament. Crown 8vo, 3s. 6d.

PLEASANT WAYS IN SCIENCE. Crown 8vo, 3s. 6d.

MYTHS AND MARVELS OF ASTRONOMY. Crown 8vo, 3s. 6d.

NATURE STUDIES. By GRANT ALLEN, A. WILSON, T. FOSTER, E. CLODD, and R. A. PROCTOR. Crown 8vo, 3s. 6d.

LEISURE READINGS. By E. CLODD, A. WILSON, T. FOSTER, A. C. RUNYARD, and R. A. PROCTOR. Crown 8vo, 3s. 6d.

STRENGTH: How to get Strong and keep Strong, with Chapters on Rowing and Swimming, Fat, Age, and the Waist. With 9 Illustrations. Crown 8vo, 2s.

REYNOLDS. EXPERIMENTAL CHEMISTRY for Junior Students. By J. EMERSON REYNOLDS, M.D., F.R.S., Professor of Chemistry, Univ. of Dublin. Fcp. 8vo, with numerous Woodcuts.

PART I.—*Introductory*, 1s. 6d. PART III.—*Metals and Allied Bodies*, 3s. 6d.
PART II.—*Non-Metals*, 2s. 6d. PART IV.—*Chemistry of Carbon Compounds*, 4s.

ROMANES.—*WORKS by* GEORGE JOHN ROMANES, *M.A., LL.D., F.R.S.*

DARWIN, AND AFTER DARWIN : an Exposition on the Darwinian Theory, and a Discussion on Post-Darwinian Questions. Part I. THE DARWINIAN THEORY. With Portrait of Darwin and 125 Illustrations. Crown 8vo, 10s. 6d. Part II. POST-DARWINIAN QUESTIONS : Heredity and Utility. With Portrait of the Author and 5 Illustrations. Crown 8vo, 10s. 6d.

AN EXAMINATION OF WEISMANNISM. Cr. 8vo, 6s.

ESSAYS. Edited by C. LLOYD MORGAN, Principal of University College, Bristol. Crown 8vo, 6s.

CONTENTS : Primitive Natural History—The Darwinian Theory of Instinct—Man and Brute—Mind in Men and Animals—Origin of Human Faculty—Mental Differences between Men and Women—What is the Object of Life?—Recreation—Hypnotism—Hydrophobia and the Muzzling Order.

SLINGO AND BROOKER. ELECTRICAL ENGINEERING FOR ELECTRIC-LIGHT ARTISANS AND STUDENTS. (Embracing those branches prescribed in the Syllabus issued by the City and Guilds Technical Institute.) By W. SLINGO and A. BROOKER. With 346 Illustrations. Crown 8vo, 12s.

SORAUER. A POPULAR TREATISE ON THE PHYSIOLOGY OF PLANTS. For the Use of Gardeners, or for Students of Horticulture and of Agriculture. By Dr. PAUL SORAUER, Director of the Experimental Station at the Royal Pomological Institute in Proskau (Silesia). Translated by F. E. WEISS, B.Sc., F.L.S., Professor of Botany at the Owens College, Manchester. With 33 Illustrations. 8vo, 9s. net.

THORPE. A DICTIONARY OF APPLIED CHEMISTRY. By T. E. THORPE, B.Sc. (Vict.), Ph.D., F.R.S., Treas. C.S., Professor of Chemistry in the Royal College of Science, London. Assisted by Eminent Contributors. To be published in 3 vols. 8vo. Vols. I. and II. £2 2s. each, Vol. III. £3 3s.

TUBEUF. DISEASES OF PLANTS DUE TO CRYPTOGAMIC PARASITES. Translated from the German of DR. CARL FREIHERR VON TUBEUF, of the University of Munich, by WILLIAM G. SMITH, B.Sc., Ph.D., Lecturer on Plant Physiology to the University of Edinburgh. With 330 Illustrations. Royal 8vo, 18s. net.

GENERAL SCIENTIFIC WORKS.

TYNDALL.—*WORKS by JOHN TYNDALL, F.R.S., &c.*
FRAGMENTS OF SCIENCE. 2 Vols. Crown 8vo, 16s.
NEW FRAGMENTS. Crown 8vo, 10s. 6d.
HEAT A MODE OF MOTION. Crown 8vo, 12s.
SOUND. With 204 Woodcuts. Crown 8vo, 10s. 6d.
RESEARCHES ON DIAMAGNETISM AND MAGNE-CRYSTALLIC ACTION, including the question of Diamagnetic Polarity. Crown 8vo, 12s.
ESSAYS ON THE FLOATING-MATTER OF THE AIR in relation to Putrefaction and Infection. With 24 Woodcuts. Crown 8vo, 7s. 6d.
LECTURES ON LIGHT, delivered in America in 1872 and 1873. With 57 Diagrams. Crown 8vo, 5s.
LESSONS IN ELECTRICITY AT THE ROYAL INSTITUTION, 1875-76. With 58 Woodcuts. Crown 8vo, 2s. 6d.
NOTES OF A COURSE OF SEVEN LECTURES ON ELECTRICAL PHENOMENA AND THEORIES, delivered at the Royal Institution. Crown 8vo, 1s. 6d.
NOTES OF A COURSE OF NINE LECTURES ON LIGHT, delivered at the Royal Institution. Crown 8vo, 1s. 6d.
FARADAY AS A DISCOVERER. Crown 8vo, 3s. 6d.
THE GLACIERS OF THE ALPS : being a Narrative of Excursions and Ascents. An Account of the Origin and Phenomena of Glaciers, and an Exposition of the Physical Principles to which they are related. With numerous Illustrations. Crown 8vo, 6s. 6d. net.

WATTS' DICTIONARY OF CHEMISTRY. Revised and entirely Re-written by H. FORSTER MORLEY, M.A., D.Sc., Fellow of, and lately Assistant-Professor of Chemistry in, University College, London ; and M. M. PATTISON MUIR, M.A., F.R.S.E., Fellow, and Prælector in Chemistry, of Gonville and Caius College, Cambridge. Assisted by Eminent Contributors. To be Published in 4 Vols. 8vo. Vols. I. & II. 42s. each. Vol. III. 50s. Vol. IV. 63s.

WEBB. CELESTIAL OBJECTS FOR COMMON TELESCOPES. By the Rev. T. W. WEBB, M.A., F.R.A.S., Vicar of Hardwick, Herefordshire. Fifth Edition, Revised and greatly Enlarged by the Rev. T. E. ESPIN, M.A., F.R.A.S. (Two Volumes.)
Vol. I. With Portrait and a Reminiscence of the Author, 2 Plates, and numerous Illustrations. Crown 8vo, 6s.
Vol. II. With Illustrations and Map of Star Spectra. Crown 8vo, 6s. 6d.

WRIGHT. OPTICAL PROJECTION : A Treatise on the Use of the Lantern in Exhibition and Scientific Demonstration. By LEWIS WRIGHT, Author of "Light : a Course of Experimental Optics." With 232 Illustrations. Crown 8vo, 6s.

www.ingramcontent.com/pod-product-compliance
Lightning Source LLC
Chambersburg PA
CBHW032048220426
43664CB00008B/916